WOOD
BENDING
MADE SIMPLE

INCLUDES
DVD
VIDEO

WOOD BENDING
MADE SIMPLE

LON SCHLEINING

The Taunton Press

The Taunton Press, Inc., 63 South Main Street, PO Box 5506, Newtown, CT 06470-5506
e-mail: tp@taunton.com

Editor: Jessica DiDonato
Copy editor: Seth Reichgott
Indexer: Lynda Stannard
Interior design: Susan Fazekas
Layout: Cathy Cassidy
Illustrator: Christopher Mills
Cover Photographer: © Randy O'Rourke
Interior Photographer: Gary Junken © The Taunton Press, Inc. except pp. 29, 30, 31, 32, 84, and 95 © Randy O'Rourke

DVD Producer: Jessica DiDonato
DVD Editing: Gary Junken

Library of Congress Cataloging-in-Publication Data

Schleining, Lon.
  Wood bending made simple / author, Lon Schleining.
     p. cm.
  ISBN 978-1-60085-249-7
  1. Wood bending--Amateurs' manuals. I. Title.

  TT200.S36 2010
  684'.08--dc22

                                    2010026529

Printed in the United States of America
10 9 8 7 6 5 4 3 2 1

introduction

Learning to bend wood opens up the possibility of adding curves to your wood-working projects. Suddenly you have the option of curved chair parts or curved table aprons. Your furniture isn't so boringly rectilinear any more.

This book is a way to get started in woodbending and introduces you to the two most practical methods—bent lamination and steam bending. Bent lamination uses thin pieces of wood glued together around a form shaped like the bend. It's probably an easier method for the first timer because it's more predictable. However, bent lamination does require the additional step of milling all the laminations. Steam bending, which uses heat to plasticize wood, has been used for ages both in furnituremaking and boatbuilding. It's tricker because of variations in temperature and timing, but that's part of the fun.

In fact, the draw of woodbending may be the challenge of convincing wood to take on new and interesting shapes. And there's an art to figuring out how to get there. There's no one right way to bend a piece of wood. Every radius, every thickness, every species bends differently. Even two pieces of wood out of the same plank might have different bending characteristics; perhaps one has a flaw and the other is clear. The idea is to approach bending as an adventure and a learning process.

Whether you do woodworking for a living or just for fun, I hope you let yourself play with these techniques. I want everyone who reads this book to experience the awe I felt the first time I measured the difference between the inside and outside surfaces of a bend. Though I've done it hundreds of times now, I'm still amazed.

Once you've mastered the basics presented here, check out my previous book on the topic, *The Complete Manual of Wood Bending: Milled, Laminated, and Steam-Bent Work* (Linden Publishing, 2002). There you'll find more in-depth information on working with curves, including sections on milling curves out of solid wood and building curved cabinet doors. Whether you make art furniture or fish nets, I hope you'll have a blast bending wood.

contents

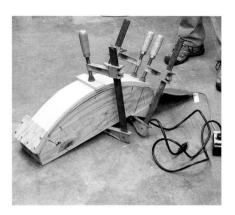

Bending forms

The bending form defines the shape of the part you wish to bend. It also holds the wood during the bending process. A good form should provide for fast clamping and a way to keep the slippery glued layers of the workpiece in alignment. For most furniture parts and other relatively narrow pieces, the form is built of layers of MDF. In this chapter, I'll show you how to draw the shape. I'll also show how to cut the layers, smooth, and join them together to make a an effective bending form.

Take your time with this step. It's very important to get the form right. Any lump or hollow in the form's curvature will telegraph into the finished part.

Computers are great for drawing, but in my opinion there's no substitute for a pencil, paper, and a big eraser. I want to be sure my clamp will fit the form. With the drawing, I'm sure.

I usually draw with pencil and paper rather than using computer drawing programs. The pencil-and-paper method provides a way for me to make subtle adjust-

Start with a drawing

Before you can make the form, you need to design the part you want to bend. Make a detailed working plan of your project, including the bent part. When you're making a bent lamination for a specific furniture part, such as a chair back, you'll need to draw a pattern for the entire project first to determine the correct size of the piece you're bending. Then make a full-size drawing of the bent part. Once you've drawn the part itself, simply drawing a rectangle around it gives you the rough shape of the form. For a simple curve, the male part of the form faces the concave side of the part.

workSmart

As you draw your project one part at a time, your brain works through the process of building it until you begin to "see" its construction in your mind. You might notice design variations to incorporate. Perhaps you'll discover a different way of joining two pieces. Most important, you might spot a mistake you were about to make, which will save time in the long run.

Materials needed to build a form

Sabersaw or bandsaw

Portable router to flush-cut the form layers

Portable surface planer

Flush-cutting router bit with bearing on the shank instead of on the end of the bit

Drill press with large drill bit to drill clamp socket holes

Cordless drill and drill bits

Brad nail gun (micro nailer) or hand nails

Edge sander or belt sander held on edge

Files, rasps, and other shaping hand tools

2-in. to 3-in. (h) × 24-in. ¾-in. plywood or MDF

MDF and/or plywood

Solid wood

Drywall screws, lag bolts

Woodworker's glue

Heavy-duty cellophane packing tape

A female form like this only makes contact in three places, distorting the clamping pressure.

ments and consider all aspects of the process. This is especially helpful for complex projects because it forces me to think about the process in manageable, bite-size pieces.

You may prefer drawing right onto a piece of MDF, but it's easier to erase changes on paper. Once you've established the final shape, you can make a clean drawing on the MDF you'll use for the form. The end product doesn't have to be a work of art, but it should provide the information you need to build the form.

Types of bending forms

Bending forms are usually referred to as "male" or "female." On a male form, or convex, form, the part bends around the form (see photo on p. 2). With a female, or concave, form (see photo above right), the part bends into the form.

One-part forms

It's easier to bend around a male form. A one-part female form creates three "hard" spots, or sharp points of contact—two points on the form and one point where the clamp contacts the bundle of laminates. Because the bundle must be pulled into the form, clamping pressure gets distorted. It's harder to pull the bent part into the correct shape. A better solution is to use a two-part form.

Two-part forms

The simplest two-part form consists of a solid male form and a solid female form. Unfortunately, it has the same drawback of using only a female form—uneven clamp pressure. The forms apply pressure in only three or four spots along the bundle of laminates (see the photo on the facing page). There is progressively less pressure as the curve of the bend increases.

Cutting the female form into sections, or clamping blocks, helps solve the problem of uneven pressure. The female clamping blocks ensure that clamp pressure is evenly distributed. The bend around the male part of the form is gradual and controlled.

To draw a two-part form with female clamping blocks, divide the female part of the form into sections. To determine the size of the sections, lay the clamps you plan to use directly on the drawing. The ideal position for each clamp is perpendicular to the curve of the part (see the drawing below.) The adjustable part of the clamp will be on the outside of the clamping block. The stationary part goes into holes in the male part. As you're drawing, lay out the locations for clamp holes.

Making a pattern

The first step in making the form layers is to transfer the drawing from paper to your pattern material. You'll be using this pattern for the top layer of the form, so mark the clamp hole locations while you're at it.

A two-part form like this one forces the part into shape very unnaturally. It can be greatly improved by simply cutting the female part of the form into sections, as shown in the drawing below.

Drawing the bending form

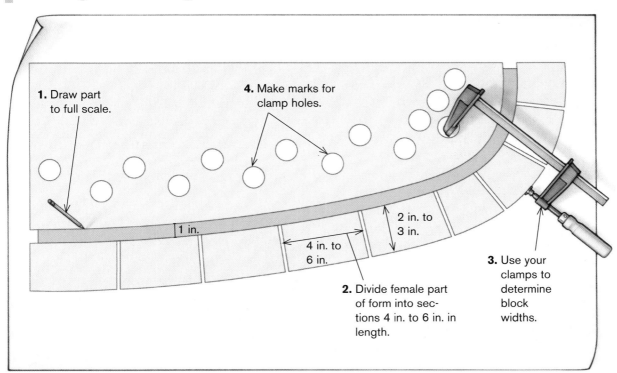

1. Draw part to full scale.

4. Make marks for clamp holes.

1 in.

4 in. to 6 in.

2 in. to 3 in.

3. Use your clamps to determine block widths.

2. Divide female part of form into sections 4 in. to 6 in. in length.

Use a jigsaw or bandsaw to cut the form layers. Cut just outside the line, then clean up the cut by hand.

It's tempting to think that making a single cut will shape the form correctly. Not so. You have to remove the thickness of the part.

The form should support the part along its entire width and length. The thickness of the form will need to be at least the width of your piece. A bit thicker is even better. Usually, you'll be using standard ¾-in. material for your form layers. For example, using three layers (a total of 2¼-in. form thickness) for a 1¾-in.-wide part will work fine.

Because the first piece of the form will serve as a pattern for the rest of the layers, it's vital to get a smooth curve. When cutting out the pattern material, be sure to leave a margin of about ⅛ in. on the waste side of the line. Yes, it leaves more material to sand or file, but if you have a little extra to work with, it gives you a greater chance of getting the curve smooth and "fair." You'll get a smoother cut with a bandsaw, but a jigsaw works as well.

If the radius is 12 in. or greater, try the router technique on p. 8 to smooth the first layer of the form. If the radius is too tight to use this technique, spend as much time as it takes to smooth the bandsaw or jigsaw cut with hand tools or sanders.

Once the pattern layer is smooth, it's simple to duplicate the shape with a router using a flush-trim bit and the first layer as a template. There are two basic types of flush-trim router bits. The more common is a straight

bit that has a bearing at the end. The other has a bearing mounted on the shank. Because your pattern is on top of the piece being cut with the shank-mounted bearing, you are able to see the process much more easily. Plus, with the bearing much closer to the router there is less vibration, which gives you a smoother cut.

Materials for the form

I like to use MDF at least for the first pattern layer because it's easy to cut and shape. MDF works well for the entire form if you're bending only a handful of parts, like legs for a single table, but it quickly deteriorates where clamp pressure is applied. In situations where you need to make many parts on the form, use MDF for the pattern and plywood for the form.

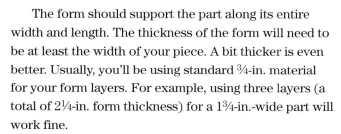

workSmart

Your fingers can detect high spots and low spots on the edge of the pattern very accurately. Mark only the high spots with a pencil. Gently sand or file only the high spots, just enough to remove the pencil marks, making sure to avoid the low spots. Repeat this technique until you're sure the edge is smooth.

Drawing the form

After you've made a plan for your project, make a fulll-size drawing of the part you wish to bend. This step allows you to work out the steps in the bending process and figure out what clamps to use and where to locate them.

1. **Draw the part exactly to size,** thickness, and shape and add extra length and width to be cut off later.

2. **Segment the female half** of the form into 4-in. to 6-in. lengths.

3. **Determine the height of the clamp blocks** based on the size of the clamps you have. Blocks should be at least 2 in. to 3 in. high. In this case my clamp is too short, and I'll need to adjust the height of the block or use a longer clamp.

4. **Locate holes for clamps** in the male form. Size the clamp holes so the stationary ends of the clamps easily fit into them.

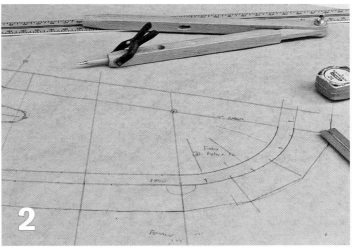

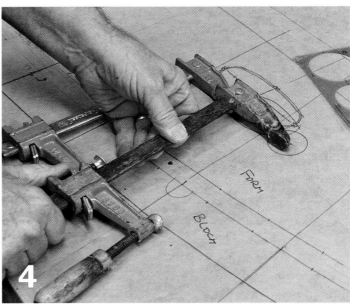

Making a pattern with a router

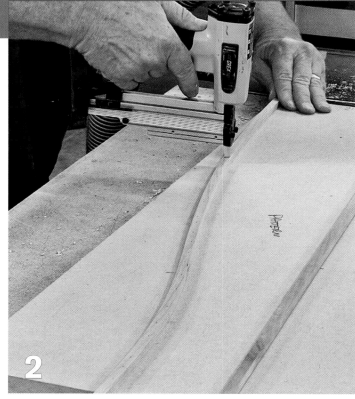

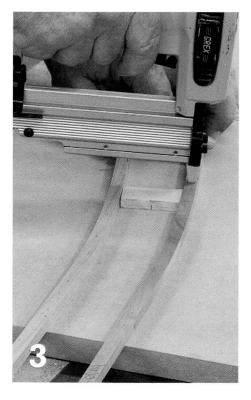

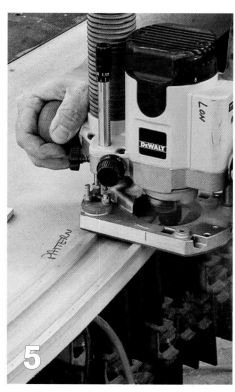

When the radius of the bend is greater than 12 in., this router technique makes short work of forming a pattern. A shank-mounted bearing flush-trim bit cuts the MDF even with the guide batten. The second batten keeps the router from tipping.

1. **Mill two identical battens** with a portable planer. The battens must be narrow enough to make the bend but exactly the same height and thickness.

2. **Pin the first batten** along the pencil line of the curve using a pneumatic brad nailer. Don't scrimp on the nails—the more the better. A very small-gauge micro-pin nailer is invaluable when nailing these very small battens because it's less likely to split the wood.

3. **Fasten the second batten** parallel to the first, about 2 in. away.

4. **Saw the waste of the pattern** to within ⅛ in. of the line.

5. **Route along the battens** to cut the pattern using a shank-bearing flush-trim bit. The bearing rides on the batten. The cutter trims the pattern material flush (see photo 3 on p. 11).

6. **Sand the routed edge** just a little to smooth any minor bumps.

7. **Remove the nails and battens**, and presto—it's a pattern. Cut the rest of the form layers using the same router and bit setup.

Smoothing a pattern cut freehand

If the bend is sharp and you can't use the router and batten method, you may have to cut the pattern freehand with a sabersaw or bandsaw. You can smooth and refine the pattern with a belt sander turned on edge. I've made this jig to support the belt sander and keep it steady.

1. **Set the belt sander** on its side.

2. **Smooth the pattern** in an edge-sanding jig.

3. **Sand out** any small imperfections in the pattern by hand.

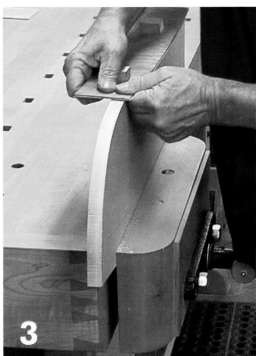

Making the layers

O nce you make the first pattern layer, it's a simple matter to duplicate the shape with a router and a flush-trim bit.

1. **Trace the pattern** on the layer blank.

2. **Use a jigsaw or bandsaw** to saw off the wood on the waste side of this line, leaving ⅛ in. to be removed with the router. Then clamp the pattern to the rough-cut layer blank.

3. **Rout along the pattern** with a shank-mounted flush-trim bit to make an identical piece.

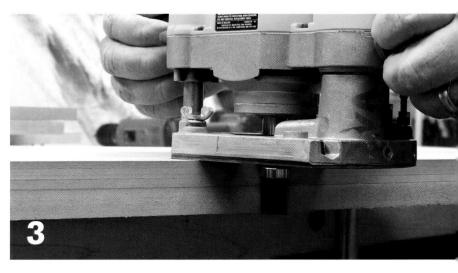

Assembling the form

1. **Apply glue** to each form layer.

2. **Screw the layers together** for extra hold. Here's a tip: If you nail them first, it keeps the pieces from sliding around while you drive the screws.

3. **Attach the guide blocks** to the form, centering them right on the gap between the clamp blocks. Guide blocks hold the bundle in place over the form during gluing. They help to keep the piece lined up on the form, providing a place for a clamp to secure to during edge-clamping and act as shelves to hold the clamp blocks steady while clamping.

4. **Drill socket holes** for the clamps using a drill press. Depending on the size of your clamps, the holes should be 1½ in. or more in diameter. Chisel a flat spot into the hole so the clamp can find a spot to grip.

5. **Place a clamping cleat** in the bench vise, put the form where you want it, mark the location, and drill holes for the lag bolts. Put the lag bolts in the form holes, and tap the bolts to mark the location of the holes to be drilled in the cleat. Then, drill smaller pilot holes in the cleat that are properly sized for the bolt threads. For these ⁵⁄₁₆-in. bolts, I drill ³⁄₈-in. holes in the form and ¼-in. holes in the cleat.

6. **Glue the joint** and tighten the cleat bolts.

3

5

4

6

Male form with female clamp blocks

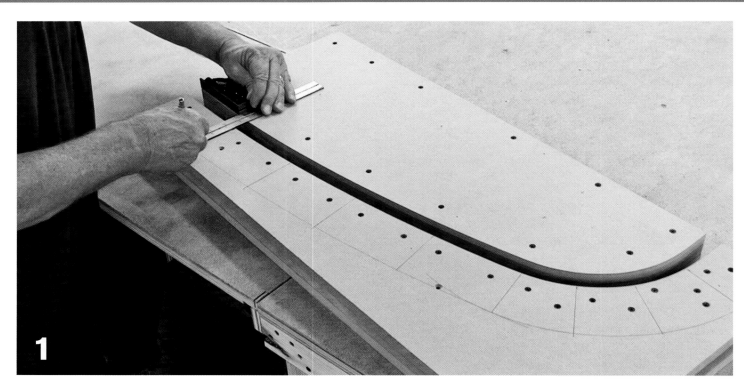

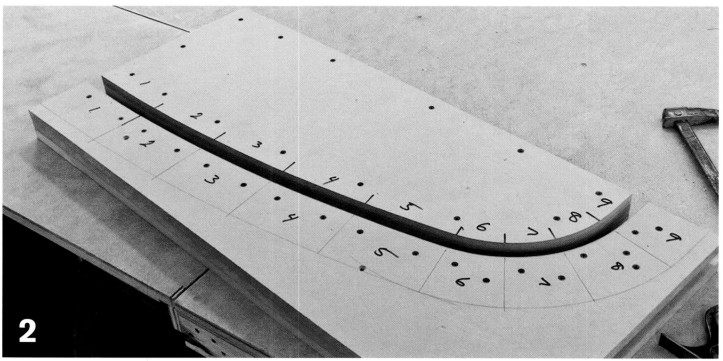

Y ou can achieve natural and gradual bends with a male form using female clamp blocks. Clamps can be oriented so they are perpendicular to the glue line, providing uniform pressure. Each section gets clamped separately.

1. **Mark each corresponding section** on male and female parts of the form.

2. **Number each section** and the form, from 1 through the ending number.

3. **Cut the female form** into sections on the bandsaw 4 in. to 6 in. long and 2 in. to 3 in. high.

4. **Cover the surface** of the form and its counterparts with cellophane tape wherever they will contact the part to keep the part from sticking to them when the glue is applied. Cellophane tape doesn't stick well to MDF edges, so you'll need to burnish it with a smooth object to affix it better.

Cutting laminates

A bent lamination consists of multiple thin wood plies glued together. After the glue cures, the plies retain the shape defined by the bending form. The plies or "laminates" need to be thin enough to bend easily around the curve. Some species are more flexible than others, so determining the thickness of the laminates may take some trial and error. Once you know the thickness, you can easily cut the laminates on the tablesaw or bandsaw.

If it takes both hands to bend a piece of wood around the curve, or if the wood feels like it's going to break, stop bending. Either plane it down a bit or cut a new, thinner piece and try again.

Determining the thickness

Deciding on the thickness of the laminates may require some trial and error because of the differences in wood species, how the wood was dried, and other factors. Begin by cutting a sample piece of the wood species you'd like to use for the bend. If the radius of the piece you're making is fairly small (6 in. or less), start with a trial piece ⅛ in. thick. If the radius is more gradual, start with a ¼ in. thickness. Make sure the sample piece is the same length as you'll need for your project. This way you'll get a more accurate feel for how much pressure it needs to bend. The test piece should be at the glue manufacturer's recommended temperature for gluing, usually room temperature. Keep in mind that these things are important because some species are more flexible than others and all tend to be stiffer when they're very cold.

Obviously, a 6-in. radius will need thinner plies than a 12-in. radius. If your test piece is too stiff to bend around the form, you'll know right away. If you have to use both hands to bend it and it feels like it will break, plane it down a little with a thickness planer. If it's still too stiff, plane it again. Eventually, it will be thin enough to bend

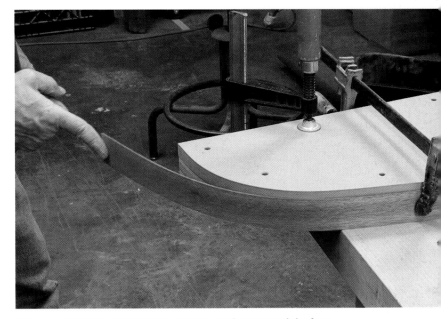

If finger pressure is all it takes to bend your piece around the form, the thickness is appropriate for bent lamination. Using this method to determine layer thickness means you don't have to allow for springback.

Determining thickness and springback

The thickness and number of laminates determine the amount of springback that will occur in a bent lamination. The goal is to keep springback to a minimum—less than ¹⁄₁₆ in.

Using trial and error to determine how many laminates to use for a bend and at what thickness drives a lot of woodworkers nuts. Engineers and others with a background in math want a formula to figure thickness. For those of you, here's a method more reliable than most.

Let Y be the value for springback. Y should be as close to zero as possible; ¹⁄₁₆ in. (0.062) or less. If the rise, or height, of the arc, X, is 6 in. and the number of layers, N, is 12, springback, Y, is minimal (0.04 in., or less than ¹⁄₁₆ in.). Let's say we're making a piece that will be 1½ in. thick. Dividing 1.5 in. by the number of layers, 12, gives us a thickness of 0.13 in., or layers just over ⅛ in. thick, to produce a finished piece with little or no springback.

Using the same math, if we make the bend using 6 layers ¼ in. thick instead of 12 layers ⅛ in. thick, we get springback of 0.17 in., or almost ³⁄₁₆ in.

When using this formula, keep two things in mind. First, there are other variables, such as the type of wood (some are more bendable than others). Second, it's important to remember this method uses the height from the cord of a given circle to its circumference, not the radius of the curve. Finally, it only applies to a section of a true circle, not a more free-form curve, which varies depending on what you need for your project.

If you use this or some other formula to mill your wood, please check to see that a single piece bends easily around the form before you get too carried away. If it's too stiff to bend with just finger pressure, it needs to be planed until it will bend easily.

Predicting springback

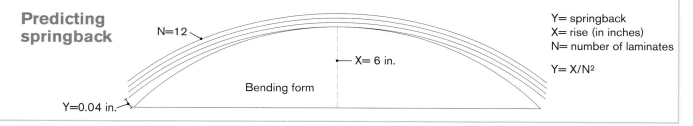

N=12

X= 6 in.

Bending form

Y=0.04 in.

Y= springback
X= rise (in inches)
N= number of laminates

$Y = X/N^2$

around the form very easily. You'll know the thickness is right when an individual piece is easy to bend around your form using only light finger pressure.

Once you figure out how thick each ply can be, it's pretty simple to find out how many you need. Just mill a length of the material to the right thickness, then cut it into short pieces. Stack the pieces up until you get the right height and count the number of pieces. That's how many laminates you'll need.

Springback

Springback happens when you release the clamps and the bent lamination "springs back" toward a straight line. A little springback of ¹⁄₁₆ in. or so is normal. The amount of springback is a function of how thick the laminates are in relation to the radius of the bend and how many laminates you use to get the thickness you want. The more layers, the thinner the laminates the less springback.

Cutting laminates on the tablesaw

You can cut laminates using either the tablesaw or bandsaw. I think it's fastest to cut them on the tablesaw, determining how much to move the fence for each cut with a jig. With the right setup and good technique, you can glue the layers fresh off of the tablesaw without having to mill them. The main limitation to using the tablesaw is that the strips cannot be wider than about 2 in., the maximum cutting height for most tablesaws. If you need wider laminates, you'll need to cut them on the bandsaw.

Mark the top of the board with a pyramid mark (see the photos on p. 21). This will help keep the strips in order. Then run the edges of the board on a jointer to make sure they are parallel. Set the tablesaw jig for the thickness of the bending pieces. This may take some practice before you get it just right. With the board held against the rip fence, move the fence to the left until the board just barely touches the bearing on the thickness jig. The fence setting may change when you lock it down, so be sure to check it before you cut. Make the rip cut as smoothly as possible. Unless the board is too long, try to make the cut in a single motion.

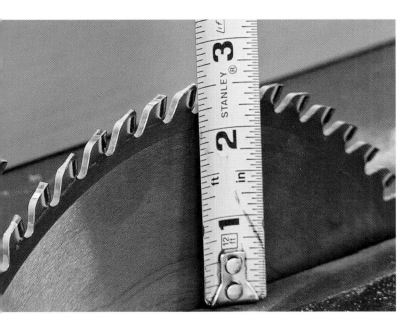

Cutting laminates to width on the tablesaw is limited by blade height. Note that on this blade, the maximum height of a cut will be 2 in. (because a cut can't be made to the top of the tooth).

Tablesaw jig for cutting laminates

You can buy or easily make a jig for ripping uniformly thick laminates on what is usually considered the waste side of the blade. Using this technique makes the operation much safer than cutting thin strips between the blade and the fence, and it sets the fence accurately to cut strips at the same thickness each time. You use the stop to set the thickness of each cut by moving the fence toward the blade until the board you're cutting hits the stop. Then you lock down the rip fence, make the cut, and repeat the process until you have all the laminates you need.

Move the cutting gauge up next to the blade to set the cut thickness. Then slide it away from the blade (toward you) and lock it down to make the cut (see the photo on p. 16).

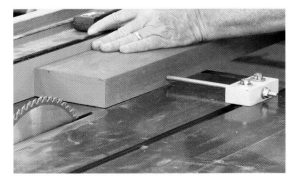

This shop-built cutting guide is quick to make, but the storebought model operates more smoothly because it has a bearing on the end to guide the cut.

Dealing with glazing or burning

Glazed or burned laminates don't bond well. Check the laminates as they come off the saw for signs of burning or glazing. Some woods are more prone to burning and glazing, particularly cherry and maple. Making clean cuts through a thick piece of hard maple takes a good, sharp blade and a saw that has enough horsepower to avoid bogging down in the cut. The saw also needs to be aligned properly and equipped with a splitter. If you see burning or glazing, just cut the pieces a little thicker, then put them through the portable surface planer (see "Planing thin laminates" on p. 22). Keep in mind that if you make the laminates thinner than you planned, you may need more to achieve the total thickness you need.

Cherry is particularly prone to burning and glazing during rip cuts. A thin-kerf blade and good technique may help remedy this problem.

Laminates directly from the tablesaw

Many woodworkers think rip cuts on the tablesaw need to be jointed or planed prior to gluing. The laminates I use go directly from the tablesaw to the glue table. All it takes is a good, adequately powered saw, a good rip blade, an outfeed table, and smooth technique for feeding the boards through the saw.

Keep the blade clean and only use it when cutting bent lamination strips or glueline rip cuts. If your tablesaw isn't quite up to the task of making these rip cuts, a high-quality thin-kerf blade might actually give you a better cut. The new blade will be very sharp and the thin kerf has another advantage. Because you're removing less material from the thinner kerf, you're using less power for the cut, boosting the effective power of the saw.

Outfeed and infeed supports

For any standard list of safety equipment I would add a stable, well-aligned outfeed table. The outfeed table supports the board so all the operator has to do is push the board through the cut. Without an outfeed table, you have to counteract the weight of the board by pushing down on its end. This is an awkward position at best. At worst, it poses a risk for dangerous kickback or pulling your hand into the blade. An infeed table is helpful to support a board while making cuts in long stock or repositioning the rip fence while using the thickness gauge.

(top) I started with two 6-in.-wide boards and cut ⅛-in. strips out of each of them, using the bandsaw for one and the tablesaw for the other. (bottom) The yield was 29 laminates using the bandsaw and 25 using the tablesaw. Note the triangle mark on the boards. This is to help keep the laminates in the same order as they were cut from the board.

Cutting laminates on the bandsaw

Many people like to use the bandsaw for ripping strips, thinking the narrower sawkerf of the bandsaw will produce more bending strips than ripping them on the tablesaw. In fact, you might get a few more pieces out of a given board. However, by the time you joint and/or plane both sides of the laminate you've cut with the bandsaw, you've removed almost as much thickness as the tablesaw kerf. The planer tends to cut more deeply on the first

2 in. to 3 in. of the laminate. That means the ends of the piece have to be cut off. Just remember to allow for this extra length if you use the bandsaw to rip the laminates.

Although I prefer cutting these layers on the tablesaw doesn't mean you should. Choose the method that best fits your equipment, experience, and comfort level.

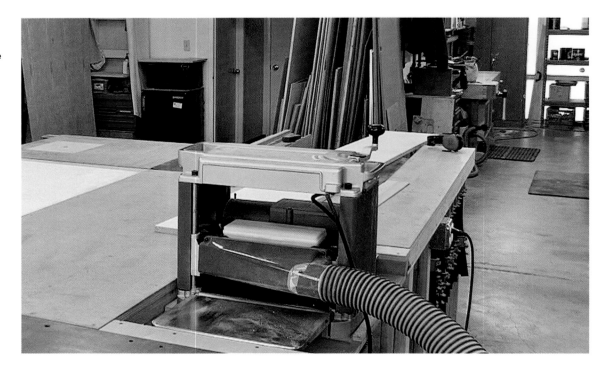

Planing thin laminates

A portable surface planer is the ideal tool to use for planing thin laminates. The key is having a flat bed without rollers, unlike large, stationary planers. Large, stationary planers are designed to mill rough sawn boards smooth. Most have bed rollers to help feed the rough boards through the machine more readily. This is why large planers aren't recommended for delicate work. Portable surface planers have no bed rollers. Instead the lower bed is solid and flat and the boards are moved through the planer by means of the top rollers.

For laminates down to ⅛ in., the setup for planing the strips is simple. Use a ¾-in. piece of melamine as wide as the planer bed to raise and lengthen the feed table. Use a screw or small cleat on one end to keep this auxiliary feed table stationary. Lube it up with TopCote®, a dry, non-silicone lubricant, then run the pieces through the planer in the normal fashion.

If the laminates need to be thinner than ⅛ in., use double-sided tape to fasten them to a piece of melamine. Then send the laminates through the planer using the melamine as a sled to hold them. You can make very thin laminates this way, down to 1/16 in. If they need to be even thinner, use presliced veneers.

I routinely plane thin strips and veneers down to 1/16 in. using my portable planer. It's important to keep the planer well tuned. I also keep a separate set of planer knives for critical planing operations like planing these laminates for bending. I often change knives; easily done with these benchtop models.

Using presliced veneers for bending

Using sliced veneers is a great option, especially when the radius is very tight and requires very thin laminates. Because the veneers are sliced with a knife, there is virtually no waste whatsoever. The nearly perfect grain match you get is possible only with these sliced veneers because there is no sawkerf as you would get if you sawed the laminates yourself. When these veneer layers are glued together, the finished part looks like a solid piece. Veneers at ⅛ in., 1/10 in., and 1/16 in. thickness are readily available from a variety of sources.

Use a strip of melamine or laminate-covered MDF to support the laminates as they go through the planer. Lube the melamine with TopCote. Don't try this with a large planer, though; it'll gobble up your thin pieces.

When the veneers are sliced with a knife, the grain match is perfect. The end grain on this bundle of matching veneers looks like a solid piece. After these laminates are bent, it will take a magnifying glass to tell they're not solid wood.

Cutting laminates on a tablesaw

1

Use a thickness guage like the one shown to cut laminates to the waste side of the blade. Before cutting, mark the board with a triangle (see the photos on p. 21 and p. 38) to aid in keeping them in the right order. The laminates should be wider than the finished dimension in your plan so there is material for scraping and planing to rough size.

1. **Move the fence** with the board held against it until it just barely touches the bearing on the jig.

2. **Make the cut smoothly,** using a single motion. If the board is too long, strategize ahead of time so that you can make the cut as smoothly as possible. Stack your laminates and tape the bundle together.

(see the photos on p. 21 and p. 38)

workSmart

If your tablesaw blade starts bogging down, it isn't necessarily dull. It might just be dirty. Pitch builds up quickly on the carbide tips. Soak the blade in some blade cleaner or household cleaner like Simple Green® and scrub it with a toothbrush. Rinse and dry the blade thoroughly. Then give the blade another try. I think you'll be amazed at the improvement in cut quality.

2

Cutting laminates on a bandsaw

A bandsaw will cut laminates almost as efficiently as a tablesaw, but you will need to plane the strips after the cut.

1. **Use your bandsaw** as an alternative to cutting strips on the tablesaw. Set up the rip fence so that it cuts a thickness slightly more than the desired final thickness of the laminate.

2. **Plane the strips** with a portable surface planer. Use a piece of melamine over the planer bed for best results. Stack your laminates and tape the bundle together.

3

Dry-clamping and choosing glue

Having the bending process go smoothly is the payoff for having prepared well. Learning to clamp quickly and efficiently is key to successful wood bending. When you're dealing with as many clamps and slippery parts as you are with most bending projects, knowing what to expect can mean the difference between success and failure. Dry-clamping is the test of your plan and your form. It also enables you to make sure everything you need is within easy reach.

Equally critical to success is choosing the right glue for the job. Most glues suitable for bending have a relatively long working time, which will help with spreading glue and clamping. But once you start to spread the glue, you need to work quickly. Being well prepared will help you work a lot faster.

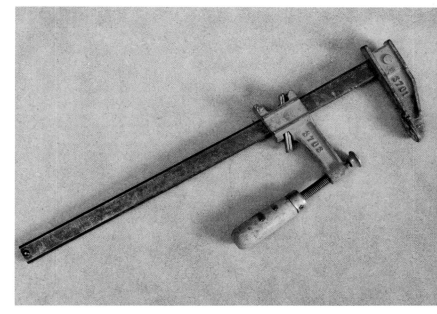

Bar clamps like this one are real workhorses. They're fast, apply positive pressure, are relatively inexpensive, and are just the right size for a lot of bends.

Why dry-clamp?

Most woodworkers quickly learn the value of dry-clamping. In a furiture project, it's the moment of truth. Do all the parts go together? Are the joints too tight or not tight enough? This stage is also an opportunity to give your clamping strategy a test run. You want to make sure you have enough clamps of the right size and you can grab them quickly.

Dry-clamping a bending project is even more crucial. Bent lamination is really just gluing and clamping. But the process must be done in an orderly, well-planned way to be successful. Although most glues for bending have a long working time, you need to spread a lot of glue. You'll also need to get the slippery laminates into the form and secure all the clamps in the right sequence. Since all bending projects will be slightly different, you'll need to get as close to the actual experience of clamping to determine the best strategy. Before you spread any glue, dry-clamp several times to be certain you've mastered clamping your project.

Bar clamps are very fast to apply and provide positive clamp pressure adjustment. They can be easily adjusted with one hand.

Choosing the right clamps

For bent lamination, I recommend using bar clamps with clutches. They're fast. They allow you to carefully adjust clamp pressure, and the clutch is very positive, making it easy to adjust with one hand. The clutch is the spring-loaded part of the clamp. It enables the screw handle arm of the clamp to slide up and down the bar. You can easily release the clutch with one hand to expand or narrow the space between the jaws.

I don't recommend hand screws, C-clamps, lever clamps (the kind you squeeze), spring clamps, or bar clamps without a clutch for bent lamination. Hand screws and C-clamps are too slow. Lever clamps are fast, but the pressure is too difficult to gauge; I never know how much I'm applying. Spring clamps usually don't provide enough pressure, and the amount of pressure isn't adjustable. Clutchless bar clamps (what I call European-style clamps) seem to have a mind of their own, especially when used vertically. They tend to slide up or down entirely on their own, making control difficult.

workSmart

There's an old joke in woodworking: How many clamps does a woodshop need? The answer is, one more than you have. A variation on this theme is you always need clamps an inch longer or greater in depth than the ones you have. If you've carefully planned your bending operation you won't have this problem. Even so, it doesn't hurt to have a few extra clamps.

Glue

Use a glue that's appropriate for bending and make sure it's fresh. Most glues have a limited shelf life. Sometimes the manufacture date is printed on the container. Carefully discard of expired glue. Don't risk wasting your time and good materials by using old glue.

Read the directions on the container. They'll have important information, including pot life, open time, clamp pressure, proper wood moisture, and working temperature ranges. Temperature is critical to proper curing. Make sure the wood, the workspace, and the equipment are within the recommended range. If your shop is large, you can make a tent from blankets so you can keep the glue, the clamps, and the clamped-up part warm enough to cure properly during cold weather.

Read the label. There's a lot of useful information in the fine print—clamp pressure, working time, open time, and curing temperature ranges.

What is glue creep?

The type of glue we use has nothing to do with springback but everything to do with "glue creep." Glue creep, or "cold creep," describes how flexible glues, like yellow wood-worker's glue, allow pieces under tension to slide or creep past one another over time as the bend tries to straighten out. This does not happen right away. It can take months or even years to show. The more tension on the bent lamination, the greater the chance of glue creep. If you've followed the guidelines for laminate thickness, you shouldn't have problems with creep, but please don't use any of the glues in the photo below.

Don't use any of these flexible glues for bent lamination. They are not designed to withstand the long-term tension of curved work.

Glue flexibility

Types of glue can be ranked in terms of flexibility. The more flexible glues are, the less suitable they are for wood bending. Bent wood is under a large amount of tension, and unless a glue is very rigid it won't hold a bend together for very long.

To illustrate the point, adhesive caulking such as 3M® 5200 glue is one of the toughest adhesives in existence. When cured, it forms an incredibly strong bond. (It's often used on boats. If you try to remove a piece of teak glued to a fiberglass hull with 3M 5200, you'll likely tear off both the wood and a layer or two of the fiberglass as

well.) Though strong, the bond is very flexible, almost rubber-like—perfect for parts that will move in relation to one another, but completely wrong for bent lamination.

White and yellow woodworker's glue fall somewhere in the middle of the flexibility scale. Yellow glue is harder (more rigid) than 3M 5200, but it's still relatively flexible. That's why it's strong enough for the all-around gluing of most woodworking projects. As the wood moves, the yellow glue stretches and shrinks just a bit so the joint doesn't fail. It's precisely this characteristic of single-part glues like Titebond® that make them perfect for general gluing, but unsuitable for bending.

These glues are suitable for bending wood and keeping it bent. Shown from left to right: Dap® Weldwood™ plastic resin glue (most recommended), West system epoxy, and Unibond 800.

"Rigid" glue

Glues suitable for bending are very rigid when cured; generally they cure hard as glass. This rigidity is essential for preventing glue creep. Take a look at the photo of hard glues on p. 31. Notice that not one of them is a single-part glue; each has to be mixed. Just associate bending with mixing glue and chances are you'll be using the right stuff.

The glue I typically use and recommend for bending is DAP®'s Weldwood Plastic Resin Glue (urea/formaldehyde glue), a brown powder you mix with water. I use it because it absolutely, positively will not creep. It is highly toxic, however, so make sure you follow the safety instructions on the label before you start your work.

Mixing plastic resin glue

Aside from the glue itself, you'll need an electric mixer and a few other items you might not have around the shop. The mixer will be permanently soiled after mixing glue. If you don't have a cheap mixer, I recommend buying one. While you're at the store, pick up some panty hose (any size or color will do), two mixing bowls (one for mixing and the other for strained glue), disposable foam paint rollers, a good supply of disposable gloves, and a dust mask.

Put enough powder into the bowl to yield about twice the amount of glue you think you'll need, knowing full well you'll be throwing away what you don't use. Mixing more than you need means that you won't have to stop

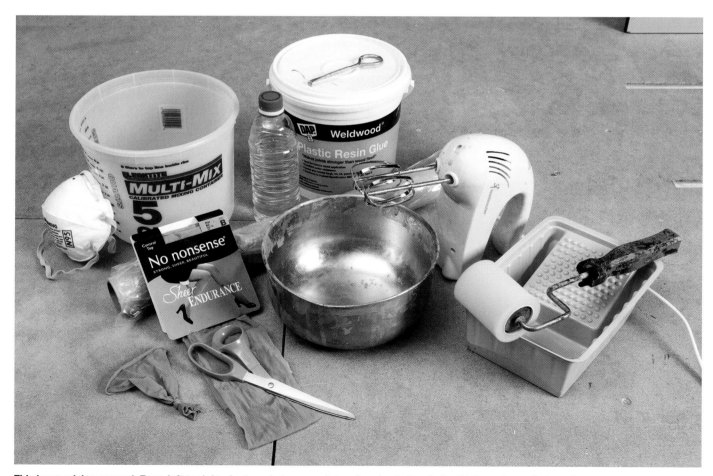

This is my mixing arsenal. From left to right: dust mask, large container for glue to be strained into, water and glue mix for making glue, panty hose for straining glue, scissors to cut panty hose, mixing bowl and mixer, paint pan and foam roller. Always wear gloves when working with glue.

Although the glue mixture will be lumpy after you first mix it, a piece of panty hose stretched over a container is perfect to strain out the lumps. The mesh of the panty hose is just fine enough for the glue to flow through while providing an effective barrier to the lumps.

in the middle of your otherwise carefully planned gluing operation to mix more. With the powder in the bowl, stir in most of the water before turning on the mixer. Just use the beaters without using the motor. This keeps the dust down considerably. Once the glue is mostly mixed, fire up the mixer on low and run it until all the dry powder is mixed into the water. Hopefully, the mixture is too thick at this point. Add water a little at a time. When the mixture is like heavy cream, that's the correct viscosity. As it gets close, add less and less water each time. If the mixture is too thin, add powder until it's the right consistency. Don't worry too much about lumps in the mixture for now. You'll strain them out with the panty hose.

warning

It's a good practice to wear a mask, respirator, *and* protective clothing when working with many glues. Always follow the safety instructions on the glue container. Some glues are extremely dangerous; they outgas toxic fumes or contain toxins that can be absorbed through the skin. Drop what you're doing and call 911 at the first sign of facial swelling or difficulty breathing.

Preparing plastic resin glue

You have the right glue (a rigid, two-part mix), you've read the label on the container, and you've followed the safety instructions. You're ready to mix glue. Remember to mix twice the amount you think you'll need so you don't run out during the glue-up.

1. **Pour most of the water** you think you need into a bowl containing the glue powder. Without turning on the mixer, stir the ingredients with the beaters until the water and glue powder begin to combine.

2. **Turn on the mixer** once the powder and water are somewhat mixed. The glue should be pretty thick at this point. If it's thicker than peanut butter, add water a little bit at a time until the consistency is like heavy cream.

3. **Test the viscosity** of the glue. It's pretty easy to add too much water toward the end. If you do, just add a bit more powder and mix well. Don't worry about lumps.

4. **Cut a 12-in. length** of panty hose including the foot, and stretch it over the rim of a paint container, bowl, or coffee can. Pour the lumpy glue mixture into it.

5. **Squeeze the glue** downward through the panty hose to strain out the lumps.

Gluing up and clamping

Clamping with glue between the layers is very different from dry-clamping. The glue acts as both a lubricant that allows the laminates to slide around and a bonding agent that quickly bonds the laminates together once pressure is applied. Once the laminates begin to bond, it is nearly impossible to adjust their position in relation to one another. The first time you experience these bending challenges you really begin to appreciate why dry-clamping is so important.

When you dry-clamp, you don't realize how much glue will drip all over the place during the actual bending process. Keep a wet rag handy to clean your gloves and clamp handles. Slippery clamp handles make it really difficult to apply the right amount of clamping pressure.

A small, disposable paint pan meant for touch-up work makes an ideal container for glue ready for spreading. The textured, angled platform allows you to roll on just the right amount of glue.

Spreading the glue

The idea of spreading all the glue necessary for a bent lamination can be daunting. There's a lot of surface area to cover before the glue begins to set. Spreading the glue quickly enough is not just a matter of hurrying. There is a system that can be used to speed the process up without really having to hurry at all.

Forget about using a brush to spread glue. You simply don't have the time. Use either a notched trowel or a paint roller. Line up the pieces according to your triangle mark. Lay them all out on plastic, side by side. Glue all the pieces on one side, then turn the pieces over, stack them up, and glue the other side. Coat only *one side* of the *outer* layers.

Once you start rolling glue, roll quickly. Don't skimp on the glue. The more you apply, the faster it will spread

and the better the quality of the glue joint. Listen as the roller runs over the laminates. As you apply more glue, the action of the roller gets quieter. When it's almost silent, you have enough glue.

Once you begin to spread glue, you're committed. You won't be able to stop until you've clamped the bundle to the form. However, if you're using plastic resin glue, it has a relatively long open time. Once mixed, the glue can sit at 70°F in the pot for up to four hours. A good routine is to set up your glue kit and workspace, mix the glue, then take a break while you mentally prepare for the upcoming bend. With plastic resin glue, you have enough time after you mix it to do one more dry-clamp if you feel the need for more practice.

When you've finished spreading glue on all the pieces, wrap the glued-up bundle with stranded packing tape in two or three spots. Make sure you can see the triangle mark and centerline on the top of the bundle.

Getting the bend started

Once you have the bundle oriented in the right direction, carry it from your gluing station to the bending form. Unfortunately, it will drip glue all over the place, which can't be avoided. If you want to protect the floor of your shop, put down a drop cloth or paper.

At first, just close the jaws of the clamps so they just hold. Applying full clamping pressure at this point won't allow you to make any adjustments in the position of the laminates in the form (see p. 40).

It can't be overemphasized how important it is to apply clamps progressively out toward the end or ends of the bend. It doesn't matter whether you start at one end and work toward the other end or start in the middle and work your way outward in both directions from the middle. In either case, the object is to gradually squeeze the glue out toward the ends of the bundle, not force it in toward the middle of the bend.

Don't force glue toward the middle

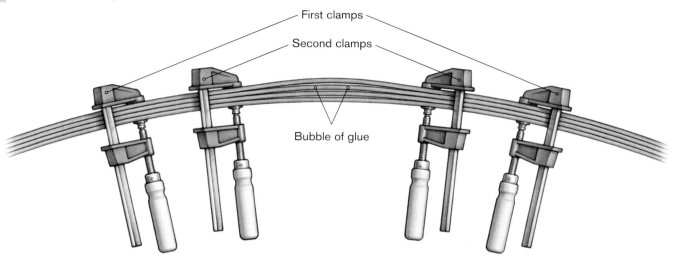

First clamps

Second clamps

Bubble of glue

Don't "leap-frog" clamps

Clamping Sequence

Creates glue bubble

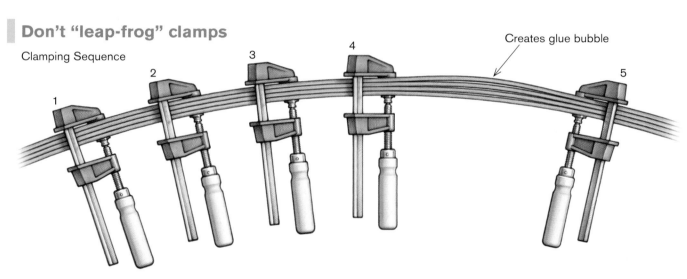

Do force glue out toward the ends

Clamping Sequence

Glue

Glue

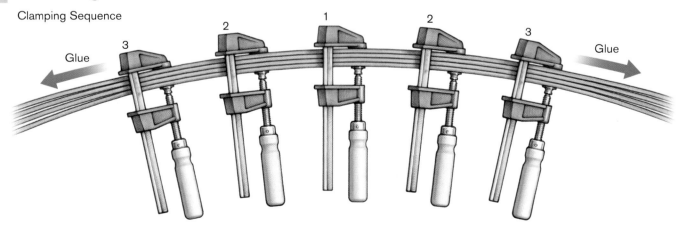

Try to remember to back away at this critical point to make sure everything is lined up. The red-handled clamp is edge-clamping the bundle, keeping it aligned.

If you apply clamps starting at the ends and work toward the middle of the bend, you force any excess glue or slack in the layers into a bubble in the middle of the bend. If this happens, there is no way to fix it except to start over. And by the time you notice the bubble, it's probably too late. Also, as you apply clamps, don't "leap-frog," or jump ahead. Don't apply a clamp and then backtrack. Apply and tighten each clamp in sequence.

Alignment

After you've loosely applied a few clamps, take a step back to confirm that they are lined up properly. Once you start applying clamping pressure, the pieces "grab" right away, so be careful to confirm that the whole bundle is in the form the way you want it before proceeding.

You want to encourage the plies to slide over one another in what amounts to a horizontal direction, but you need to carefully control them in a vertical direction to keep the bundle intact. Do this by edge-clamping (see the photo above) . When you built the form you added cleats for this purpose (see step 3 of "Assembling the form" on

p. 12). The cleats are intended to give you something to which you can apply the vertical edge clamps. Otherwise, the laminates could easily get misaligned as you clamp.

To correct misalignment you must attempt to un-clamp and then re-clamp the entire piece. It isn't usually possible to do this quickly enough before the glue starts to grab. It's usually better to throw the part away if you have trouble clamping. Don't even try to scrape the glue off of the pieces. You'll never remove it all. Any hardened glue will prevent the laminate from meeting properly with its neighbor. Use fresh laminates instead and consider what you can do differently to make the bend go more smoothly the next time.

Making the bend

Begin by loosely applying a single clamp in the center, then add a single clamp at each end, again loosely. Check the laminates' alignment and apply edge clamps to keep the laminates in line. Then—and only then—apply clamps, either starting in the center and working out toward the ends or starting at one end and working out toward the other. Tighten the clamps as you go.

Apply clamps progressively out toward the ends of the piece, using your clamp blocks to distribute the clamping pressure. Get the clamps as tight as can be as you go. Don't backtrack to apply more clamps or retighten a section already clamped.

The last thing to do after the clamps are in place is to take a stick or your gloved finger and scoop off as much excess glue as you can while it's still liquid. Glue is a lot easier to remove wet than after it's cured.

Don't forget check the temperature in the shop to make sure it's within the range for the glue to cure properly. If it's going to get too cold (below 65°F) in your shop overnight, it's pretty easy to make some sort of heated enclosure. I normally use sawhorses and moving blankets to make a tent. You can place a small space heater or even a light bulb inside the tent to keep everything warm until the next day. Avoid heating setups that could pose a fire hazard such as an open-flame heater or heating coils too close to fabric.

Get the clamps as tight as they can be as you go, working out toward the ends on each side of the center. It's OK to check to make sure they're tight, but don't try to adjust them.

Notice the vertical clamp with the red handle. By applying pressure in this direction, not only are the laminates forced into position so they line up with their neighbors, but the entire bundle stays straight on the form just like you want it.

Remember to remove as much excess glue as possible while it's still liquid.

Gluing up the laminations

1. **Cover the bench** with plastic or paper. Tape the plastic down so it stays put.

2. **Mark the bundle** with a triangle mark using a Sharpie®. A triangle mark on the top of the bundle makes it easy to tell which is top and bottom, what the sequence of pieces is, and if any of the laminates are switched end for end. Do not be concerned about removing the mark, because later you'll mill this surface.

3. **Clean the dust** off of every square inch of glue surface with either a vacuum or compressed air.

4. **Line up the laminates** edge to edge. Be careful to place the two outside pieces face down on the plastic. This sets up the entire bundle to coat one side with glue.

5. **Spread glue over** the entire surface with a disposable paint roller. Keep dipping the roller to add glue until the rolling action gets very quiet.

6. **Stack the laminates glue side** to glue side once you've covered one side of the pieces. Flip them over and move them again so the edges touch. Remember, the outside layers get glue on only one side. Repeat the process on the other side until you have the whole dripping bundle of glued strips coated. Check the triangle mark one more time to make sure the pieces are in the right sequence.

7. **Use stranded packing tape** at intervals along the bundle to keep it together for clamping. Taping reduces the very limited time glue has before it skins over if left exposed to the air.

8. **Carry the bundle** over to the form. This is where wrapping the bundle with packing tape really helps control it until you get it on the form and ready for clamping.

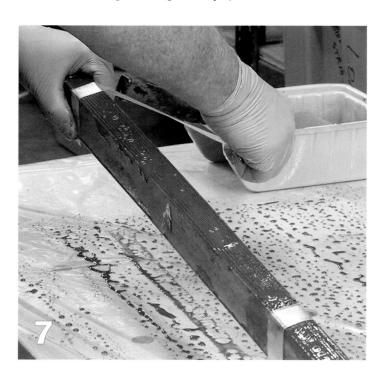

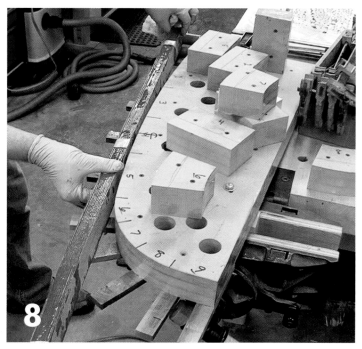

Bending and clamping the glued part

1. **Place a drop cloth or paper** underneath the table to catch glue drips.

2. **Line up the center mark** on the bundle with the center mark on the form.

3. **Place a clamp** in the center, making sure the bundle layers are even with one another. Get the clamp snug but not completely tight quite yet. Apply a vertical clamp with a block to make sure the layers stay even and flat across the top.

4. **Pull in and loosely clamp** one end, then the other. This gets the piece started bending but still allows the pieces to freely slide on one another, the key to a good bend. If the part is aligned properly, apply clamps and blocks to the left and right of the center block, tightening the clamps just snug. Then tighten the center clamp.

5. **Apply vertical clamps** and blocks as you go to make sure the layers stay aligned with each other.

6. **Apply clamps and blocks to the left and right** of the three blocks you have in place. Once you have these clamps snug, go back and tighten the clamps on the previous blocks. Continue applying blocks and clamps out toward the ends until you have all the blocks in place and all the clamps tightened.

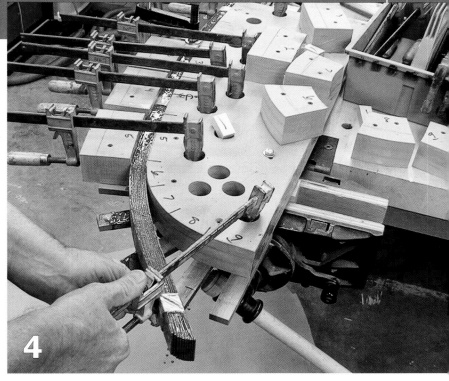

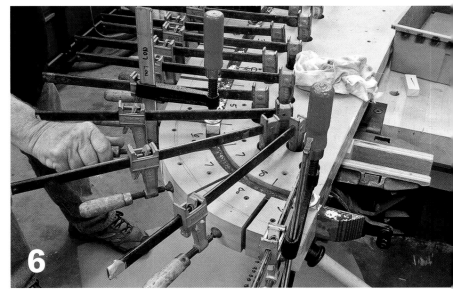

Milling the bent lamination

I t's the next day. You're anxious to see the results of your glue-up. You can hardly see the part amid all the clamps, blocks, and excess glue. As you release the clamps, you hear an unsettling crack once in a while. Not to worry. During the clamping, the part may have shifted just a bit.

Loosen all of the clamps and remove them to expose your bent lamination. You may have to tap the part just a little to get it loose. Without moving it, check for springback. As you should see, there is barely enough springback to slip a credit card between the part and the form. The bad news at this stage is the part is covered with glue. The good news is that removing all the excess glue will be a lot easier than you think.

Tap the piece lightly to make sure it's broken loose from the form. You can instantly see if there is any springback. The springback with this bend is less than 1/16 in.

warning

This glue is not only hard as glass, but it's sharp as glass too. It will slice your finger before you know it. Wear leather gloves when removing the part from the form and keep a first-aid kit on hand. Please use only the method I explain to mill your bent lamination. Milling a wooden piece of this shape on a tablesaw or a jointer invites disaster.

Removing the glue

Before scraping any glue, use a pencil to renew the center mark on the bent part. As you scrape and sand to remove glue from the part, keep renewing this center mark. It will be very useful as you build your project.

The goal of scraping is to remove as much glue as possible ahead of planing and sanding. Hardened glue is hard on planer knives and abrasives. Use a hook scraper

Use a hook scraper to remove most of the glue. If you keep it really sharp, it works amazingly fast.

filed to a razor-sharp edge. It will take off most of the protruding lumps of glue. Three things to remember about using this tool:

1. A hook scraper requres two hands.

2. Work slowly and carefully, especially until you get used to it. Otherwise, you may cut too deeply, gouging your nice curved part.

3. You need to sharpen the scraper frequently.

Sharpening your scraper

The blade should be plain steel as opposed to carbide. You can file plain steel to a much sharper edge than you can with carbide. Sharpening is quick and easy. Sharpen the scraper like you would a chisel, to a beveled edge. I normally rest the scraper on the edge of the bench, then file downward (see the photo above).

Scraping is a two-handed task, so the bent lamination must be secured to the bench before you begin.

Scraping glue

There will be very sharp shards of glue where the piece contacted the form, so don't scrape so hard that your hand slides over a jagged area. The part needs to be securely clamped to the bench so you can use both hands to work the scraper. Pull and steer with one hand, and apply light pressure with the other. Remember the scraper will dull after only a couple dozen cuts and then you have to refile it. Give it three or four swipes with the file and you're back in business.

Sanding the edges

After you've scraped off most of the glue, you can lightly sand the edges with a belt sander using a vacuum attachment to pull the sawdust out of the air. You're not looking for a finished product at this stage, just relatively smooth edges. Let the faces of the part go for right now and concentrate on the laminated edges.

Dust from the glue can be very toxic. By hooking the belt sander up to a shop vacuum with a very fine filter, most of the dust is sucked into the vacuum.

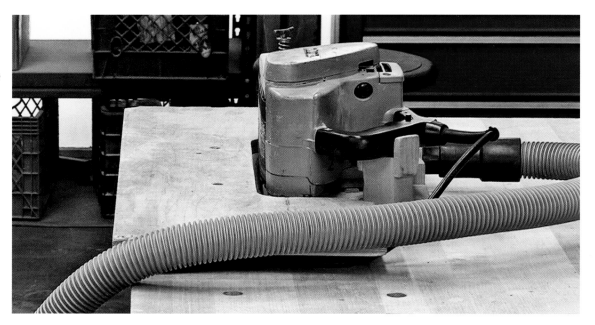

This belt sander and temporary jig works well for sanding the surface of a bent lamination, but using an edge sander is also a viable option.

A microfilter shop vacuum is hooked up to the belt sander to aid in dust collection.

With small curved parts like this one, it's best to use a stationary belt sander or to make a fixture to hold a portable belt sander on its side. With larger laminates it's easier to clamp the part and then sand it with a portable sander. The idea is to get the edges as flat and free from excess glue as possible just to make a suitable surface to run through the portable surface planer. It's the planer that does the real work.

Planing the part

Working the edge of a bent lamination is very much like milling the edge of a piece of plywood. It's simply a fact of life that the glue is very hard; hard enough to damage your edge tools. Keep that well-used set of knives for your surface planer around just for jobs like this. You can swap out the knives in most planers in about five minutes.

This is a technique for small portable surface planers only. If your planer is too heavy to lift by yourself, it's probably too big for this operation. Do not try it on a large stationary floor-model planer because it will likely make too powerful of a cut, ruining your bent lamination.

Planing to rough size

When you have both edges reasonably flat from scraping and sanding, put the flattest edge down onto the planer

This is about how far to take the scraping and sanding. A few glue spots show through, but the piece is relatively smooth on both sides. Put the best side down on the bed of the portable surface planer and the worst side up toward the knives.

bed for the first pass through the planer. Having the less flat edge up toward the knives means the planer will flatten this edge. Make several very light cuts, around ¹⁄₆₄ in., taking off just a small amount with each pass. A quarter turn at a time on the thickness adjustment is plenty.

Once the upper edge is flat, flip the part over, give the adjustment another quarter turn, and plane the other edge. Alternate edges until both are flat and parallel. You want the part to be cut with the grain as much as possible, so it's a matter of steering the piece so it stays in the center of the planer as it goes through.

Planing to final size

If you're making more than one bent lamination, wait until you're finished bending, sanding, and surface planing all of them before planing to final size. At this point all of the bent laminations should be the same width, with

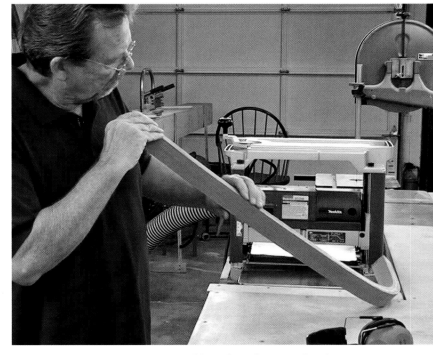

Make very light cuts through the portable surface planer, turning the part over for each pass.

It's easy to achieve a smooth, good looking edge surface on a bent lamination when using the planer.

flat and parallel edges, and just slightly wider than the width your design calls for. Pass the bent parts through the planer, one after another, as many times as necessary to reach your final dimension.

Cleaning the glue off of the face

Scrape the glue off with the hook scraper first. A sharp scraper will clean most of the glue off of the faces of bent laminations. Be careful to keep the surface as smooth as possible, but scrape off as much glue as you possibly can.

Finish up the convex (outside) face with the belt sander on edge. If you don't have a stationary edge sander, you'll find that a large 4-in. by 24-in. sander held on edge works very well. Use 120-grit belts. You'll be amazed how fast it will clean the face.

The inside, or concave, face is a bit trickier to sand. Use an insert like the one shown in the photo at right to sand a gradual radius. The insert is simply rounded piece

workSmart

When you cut the laminates, you purposely made them both wider and longer than your finish part size. When you clamped, you were careful to edge-clamp to keep the laminates lined up with one another. You did this for a reason: to give yourself plenty of extra material to plane down to your finish size. If you had not made the laminates wider than finish size, you would either have to stop planing before the excess glue was completely gone or mill the part undersize.

The scraper does a truly remarkable job of removing glue from the face side of the bent lamination. Sharpen the scraper blade frequently for best results.

To easily sand the inside (concave) face of the bent lamination, place a rounded wooden insert under the cork pad at the base of the belt sander. The tension of the sanding belt keeps it in place while you smooth out the curve.

of wood as wide as the belt. It fits under the belt and under the sander's flexible platen to make the surface on the sander curved (convex) so you can sand the inner (concave) part of the radius more easily. To make the insert, begin by cutting a ¾-in.-thick block the width and length of the sander's flat surface. Then cut a radius as smooth as possible on the bandsaw. Release the tension on the belt, place the insert under the cork pad on the sander, and away you go. If the radius is too tight, use the front wheel on the belt sander or a spindle sander to finish the part.

You may need a little hand sanding at the end of the process. If you machine sand with 120-grit, begin hand sanding with 100-grit. Then hand sand with 120-grit. Make sure you sand very thoroughly. It's easy for a shadow of glue to remain unseen until you brush stain or sealer onto the piece.

This finished chair arm illustrates how difficult it is to distinguish solid wood from a bent lamination if it is done well.

workSmart

Make sure the part is smooth, especially if you're going to round the edges with a router. The router bit is guided by a bearing riding on the surface of the part. If the surface is rough, this roughness telegraphs to the router cut.

Milling the bent lamination

After the clamps are removed, the piece will need to be scraped and sanded to remove excess hardened glue squeeze-out and planed down to the desired size.

1. **Scrape the glue** with a sharp hook scraper. The blade should be plain steel so that it is easy to sharpen. This is essential, as you will be sharpening the blade often.

2. **Lightly sand the edges** with a stationary belt sander to smooth them. Hook the sander up to a shop vacuum to aid in dust collection.

3. **Feed the bent lamination** through a portable surface planer, guiding the workpiece so it doesn't get stuck. After a light pass or two on one side, flip it over for another couple of light passes. Take off no more than a quarter turn at a time (a quarter turn equals a cut of about 0.02 in. on my planer) with each pass. If you're making multiple bent laminations, wait until they are all complete and surface-prepped and then make as many passes as necessary to meet the correct final dimensions.

4. **Sand the face** to remove any residual glue, then belt-sand the outside (concave side) of the part with a 120-grit belt to buff out the remainder.

5. **Place an insert** into the belt sander and run the convex side (inside) of the part around it. Hand-sand with 100-grit and then 120-grit sandpaper to remove any glue that may be left over.

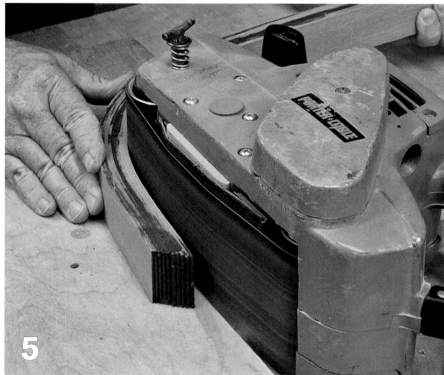

Setting up for steam-bending

Steam-bending is the process of using steam heat to make wood smalleable enough to bend. After bending, the wood dries and cools around a form, rehardening in its new shape. This sounds easy enough, but in reality the likelihood that a given bend will work depends on controlling lots of variables.

We can manage factors such as wood species, straightness of the grain, steambox temperature, steaming duration, and applying enough end pressure. The right steam-bending setup will ensure that you have a good handle on these basics. From there, as you experiment, the experience you gain will help you get more predictable results. You'll learn something from every bend.

I usually build a new box for each project and make it just big enough for that project's parts.

Making a steambox

You don't need an elaborate setup for steam-bending. Anything that will boil water will generate steam. You'll need to channel that steam into a ventilated container (usually a wooden box), where you can place the wood you want to bend: the steambox.

A steambox must be large enough for the project and the boiler strong enough to heat the box to a minimum of 200°F for the entire time the wood is in the steam. The bigger the steambox, the more steam it requires to bring it up to temperature. It stands to reason that the larger the box, the larger the boiler or number of boilers you'll need to heat it.

Unless you plan to do production work, try not to think of steamboxes in terms of one size fits all. Instead, make them just the right size for a particular project. For example, instead of heating 12-in.-long pieces of oak in a 5-ft. steambox and using your largest boiler, you can make a steambox to fit your pieces and use a tea kettle as a boiler.

Use coated deck screws designed for exterior use when constructing your steambox. Regular steel materials will leave a black stain on wood and can ruin your workpiece.

When you build a steambox, make shelves to keep some space between your bending pieces. (I like to use dowels.) This lets steam flow around the pieces freely, heating them evenly.

Materials

Exterior plywood or solid pine works great for steamboxes. Pine is relatively inexpensive and comes in a variety of lengths. If you're bending a piece longer than 8 ft., the long lengths that are available make building a 10-ft.- or 12-ft.-long box much simpler.

After you've cut the box-making materials to the size needed for your project, fasten them together with screws. Use coated deck screws designed for exterior use instead of using steel nails or screws. Steel and steaming hot wood are a poor mix. If steel comes in con-tact with wet wood—especially oak—it will leave a very deep, hard-to-remove, black stain.

Shelves in the steam box allow the steam to circulate around the wood to be bent. An easy-to-build shelving system is rows of wooden dowels that go all the way through the box. Drill the holes for the dowels before assembling the box. Make sure the holes are high enough to suspend the bending planks above the floor of the box and wide enough apart from each other that steam can circulate on all sides.

A simple cooking thermometer makes it very easy to measure how hot the box is. Calibrate the thermometer by dipping it in boiling water. If you're near sea level, the thermometer and the interior of the steambox should read 212°F.

To make the door, cut a piece of wood to the right size so that it can hold it in place with a friction fit. Add a handle cut on the bandsaw and just screw it to the door.

Rigid foam is a good way to insulate the steambox if the temperature gets close to 212°F, but not quite as hot as needed. Insulating the box is a good alternative to adding another boiler. Rigid foam is readily available at home centers. Cut it to size with a knife and tape it in place around the box.

Conducting and venting the steam

The best way to get the steam into the box is by means of high-temperature rubber hose, which is widely available in a variety of diameters. While you're at the store buying the hose, buy a drill bit that corresponds to the outside diameter of the hose. It's a simple matter to drill a hole in the side of the box to the size of the hose. Cut the hose to length and place it in the hole.

Vent holes in the steambox allow proper steam circulation and keep pressure from building up inside the box. Too much steam pressure in the box can cause an explosion. Place a meat thermometer in one of the vent holes to read the temperature inside the box.

warning

Sometimes it's hard to find vent holes. Circling them with a marker—before turning on the boiler—makes them more obvious, so that you can avoid being scalded by extremely hot (and invisible!) escaping steam.

Using PVC pipe as a steambox

It seems everyone, including myself, tries using PVC as a steambox at least once. It does work in a pinch, and it's quick to set up. The trouble with PVC is that when the pipe gets hot, it sags. The work-around for this is to make a "V" out of wood or plywood as long as the pipe for support.

If you decide to try using PVC as a steambox, be sure to drill some holes for dowels every couple of feet. Insert the dowels for shelves to keep the piece off of the bottom of the pipe. Otherwise, the steam won't circulate evenly.

Lots of people use plastic pipe for steam-bending boxes. Be mindful that it tends to sag when it gets hot.

Building a "V" shape out of lumber will support the plastic pipe and keep it from sagging.

Here are some of the boilers I use most. From left to right: A large, homemade boiler, two electric kettles, and a turkey fryer.

Generating steam

There is a definite relationship between the size of the box and the size of the boiler. It's absolutely essential that enough steam is generated to heat the interior of the box to at least 200°F and to keep it at that temperature for the duration of the heating phase of the bend.

While it's possible to calculate the size of the boiler you'll need, it's easier to build the box first; so make a guess as to the size of the boiler, and then give it a try. If the box promptly heats up to 212°F, you've got the right size boiler. If not, add boiler capacity until it does. Raising the temperature is literally a matter of adding more steam. You can gang boilers to generate more steam. For example, if using one steam kettle gets the box to 150°F, chances are two kettles will get it to 212°F. For a larger box, you'll probably need a boiler that produces more steam.

This large, two-burner boiler really puts out the steam. I can run just about as large a steambox as I could imagine using and still get it hot with this setup.

Tea kettles put out an amazing amount of steam for their small size.

It's pretty easy to build a lid for the turkey fryer that will channel steam into a hose. I used plywood, some weather stripping, and plastic plumbing parts.

Tea kettles

Tea kettles work very well for heating small steamboxes. They create more steam than one would expect, they boil water quickly, and because they're electric, they're among the safest to use of all boilers. A regular tea kettle will be sufficient to use, along with a large diameter rubber hose to funnel the steam into the box.

Turkey fryer

Many people use the popular setup for deep-frying a turkey as a boiler for steam-bending. The burner and boiler are preassembled and emit plenty of BTUs to heat up even a large steambox. The biggest challenge with using a turkey fryer is adapting the lid to funnel the steam through a hose and up into the box. Automotive heater hose, which is easy to find at any auto-parts store, works best for this function because it can withstand high heat. If the hose is not designed for high temperature, it will probably collapse and may even melt.

Many turkey fryers use propane as a source of heat. Any boiler setup using an open flame to generate heat should only be used outdoors.

A turkey fryer is a great setup for boiling lots of water because it can put out a large amount of heat.

Automotive heater hose is easy to find at auto-parts stores. It makes funneling steam from the boiler to the box pretty easy with the help of an assortment of pipe fittings and hose clamps.

warning

When you're cobbling something together, such as a lid for a turkey fryer and automotive heater hose, please make sure it's as safe as possible. If something comes loose or pops off, someone could be seriously burned.

The form for steam-bending (right) is similar to the one you use for bent lamination (left). The two major differences are that the steam-bending form dramatically over-bends the shape and it has a compression strap built onto it to facilitate bending.

Steam-bending forms

Forms for steam-bending are similar to those used for bent lamination, but are different in shape and strength because you'll be over-bending the wood and using more force to do so. Because of the tension on the form and moisture from the steam, you'll get more life out of a steam-bending form made from plywood than one made from MDF.

Making the steam-bending form smooth is just as important as it is for bent-lamination forms. As with any bending operation using a form, a lump or hollow in the form will telegraph into the finished piece.

workSmart

Springback is a fact of life with steam-bending. The key to dealing with springback is to over-bend. Too much over-bend is much better than too little. Assume the piece is going to spring back 15% to 20%, maybe more. Make the form accordingly so that you'll over-bend the piece by this amount.

Trial bending

Bending just for practice is a great way to get a feel for the process. Bend enough samples to feel confident when you approach an actual project. No one can tell you exactly how fast or slow to work when bending a particular piece of wood. You'll learn it more easily if you experiment. There are many factors affecting a bend, including the wood species, the moisture content of the stock, and the straightness of the grain. Flaws will also affect how well the bend will work. So include stock with flaws in your trials to see how. This will also teach you the importance of good stock selection.

Trial bending is a good way to try to predict how difficult or easy your bend is going to be for your particular project. Before you spend time and money setting up to steam-bend the part for your project, you may end up deciding that steam-bending isn't predictable enough. Remember that bent lamination is also an option.

Determining moisture content and why it's important

If you try to build furniture when the wood moisture content (MC) is too high (above 7% to 10%), the wood will shrink as it dries. Wood with a high MC is unsuitable for building furniture because it won't hold a glue joint, and any joinery you carefully fit will change as the wood shrinks. And the wood won't hold a finish either. (For additional information, please read Bruce Hoadley's *Understanding Wood.*) So it pays to make sure the MC of the wood that you're using for your project is within the acceptable range of 7% to 10%.

Here's a low-tech way to determine moisture content: weigh a sample of the wood species you're using, record the weight, and dry the wood in the oven. Remove the wood and weigh it again. Repeat the process until the wood remains the same weight after several times in the oven. Divide its final weight by its original weight to determine its percent of moisture content.

Many otherwise reliable sources on bending are adamant that MC must be higher than the normal 7% to 10% to

Weigh a sample of wood and then dry it. Divide its final weight by its original weight to determine percent of moisture content. The optimal percentage of moisture content for building furniture is 7% to 10%.

bend successfully using steam heat. They sincerely believe you have to use uncured or "green" wood, or at the very least wood that's been air-dried to some higher-than-normal MC. These are often the same sources that suggest you must soak wood to bend it, or that MC must be at some level or other to make certain bends. They say you're wasting your time trying to bend kiln-dried (KD) wood because it's "case-hardened," or too dry for bending. These beliefs are

inaccurate. Even kiln-dried wood steam-bends very well at 7% to 10% MC.

The bottom line with bending wood using steam heat is you can use whatever wood you have available. With the right setup, you can bend wood regardless of MC, and whether or not the wood is green, or has been soaked or air-dried. What is important is that you get the wood to an MC of 7% to 10% before you try to build furniture out of it.

Steam-bending safety

Before you start firing up your boiler to bend wood, here are some safety considerations. Most are obvious, but sometimes the most obvious things are the hardest to remember.

- Open the door to the steambox very carefully. Let the steam escape for a few seconds, then remove the piece you're bending.

- Wear welding gloves that protect your forearms. Reaching into the steambox is roughly equivalent to reaching into boiling water.

- Continuously monitor the water level in the boiler. You will go through water faster than you think. You shouldn't let the boiler run dry, but if it does, do not add water until it cools. Adding cold water to an empty hot boiler could cause an explosion. Let the boiler cool completely and start over.

- Don't use an open flame inside your shop if it's avoidable. An electric heat source is a much safer alternative.

- Don't run steam under pressure, either in the boiler or the steambox. Drill holes in the steambox to let the steam vent. Steam under pressure is potentially explosive.

Stock for trial bending

Cut a variety of pieces all the same length and beginning at about ½ in. sq. up to ¾ in. sq., or larger if you want. Try oak, pine, mahogany, or whatever you have around the shop. If you have some scraps of molding around the shop, now is a great time to test bending some different configurations. The larger your variety of species, sizes, and molding profiles, the more useful this exercise will be.

Load pieces into box

Load your trial bending wood pieces into the box while it's still cold. Spread them out so steam can circulate around them. Heat the box and measure the temperature with a meat thermometer. The box should get up to 212°F in about 10 minutes from the time you plug in your boiler if you're using the proper setup. Write down the time when the box reaches full temperature.

Heat ½-in.-sq. samples for about 30 minutes. This is based on the general guideline to heat the pieces one hour per inch of thickness. Note that this duration is only one of the dozens of variables with steam-bending and one of the few you can actually control.

Replenishing the water supply

Chances are you'll need to make steam for several hours. This means you'll have to replenish the water supply in the boiler. In the case of a tea kettle, it's easy to check it periodically to see if it feels like it's running low because of the change in weight.

Adding more water can be a very tricky operation, as it will have to be added while the boiler is hot. If you're using a kettle, heat water in an extra tea kettle to boiling and make a setup like the one shown at right to safely add water to the kettle as it is still producing steam. This way the boiler never cools down during the replenishment process. Never let the boiler run dry. If it does, you'll need to cool it down completely and start over to avoid an explosion.

Here's a good way to add more hot water to the boiler when it gets low. A small funnel with a piece of hose attached is tacked to the steambox. The hose leads to a hole in the pipe attachment of the electric kettle, where the water enters the boiler. Refilling the kettle this way allows continuous steam to enter the box, ensuring the wood pieces inside don't start to cool in between fill-ups.

warning

Do not let the boiler run dry. In the event that it does, let it cool completely before adding water and starting over. Adding cold water to a hot boiler could cause an explosion!

Make the bend

Wearing gloves, take out a piece and test it. Note that it isn't even wet; nor does it feel particularly flexible. See how far the piece will bend. Try bending it slowly, more quickly, around a gentle radius, then around a sharper one. After about ten minutes, take another test piece of the same species and try to bend it. See if the extra time in the box makes any discernible difference in how the wood bends. Try this with samples of different species and note how well or how poorly each species bends.

You should also see that as the radius gets smaller, the piece gets thicker, or more of the stock goes around the bend and becomes more difficult to make the bend.

Bending around a form

After you've bent trial pieces and selected the best stock for your project, you can move on to bending your actual workpiece. You will use two forms: a bending form and a cooling, or setting, form. The former is for making the initial bend, during which the piece is over-bent to prepare for springback, and the latter is the form that should be the exact size and shape you'd like your bent wood to be to fit your project. Your piece stays on the bending form for only a few minutes before you transfer it to the cooling form.

Bending form

When the wood has been steamed at 212°F for the appropriate time (about one hour per inch of thickness), it must be removed from the steam and immediately brought over to the bending form. The bending form must be bolted or clamped to your workbench so it won't move around while you bend.

Any time you steam-bend a piece thicker than about ½ in. or a radius smaller than about 3 ft., you'll need a compression strap (see Chapter 7). End blocks on the strap prevent excess stretching as the piece bends. (Stretching is the chief reason bends fail.) You place the strap on the piece you're bending, apply a moderate amount of end pressure, and then bend the piece over the form, clamping as you go. Once it's in place, you need only leave it on the bending form for 10 to 15 minutes,

just long enough for it to cool just a little. Then remove the piece from the form, take it out of the compression strap, and clamp it to the cooling form.

Cooling form

The cooling form is where the wood will cool after it's been bent and sprung back. The form should be built to the exact shape of your finished piece. Clamp the piece to the form and leave it to cool, dry, and set to the correct shape. To help the process, place the wood and cooling form on a rack above a space heater and cover with a blanket. If you happen to have a sauna, that would be the perfect place to put it. The bend should be completely set within 24 hours.

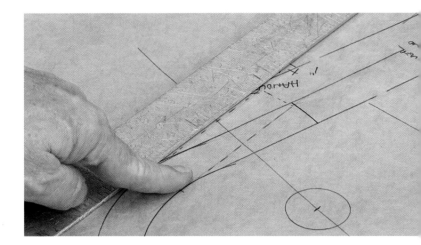

Plan for springback by building a bending form that will ensure the piece is over-bent. Make the cooling form the exact shape of the finished workpiece.

FLEXIBILITY

The purpose of this chart is to help guide your decision as to what bending method to use for a given species. The radius, degree of arc, and thickness of the piece all combine to make this a severe bend, which is nearly impossible to perform without compression, regardless of species. I deliberately chose to test species on a bend like this in an attempt to isolate a single variable with steam-bending: the type of wood.

Samples were all purchased as surfaced kiln-dried lumber. They were each milled to 1 in. square, then steamed for 1 to 1½ hrs. Each was bent with a compression strap 90+ degrees over a 6-in. radius. Ten pieces of each species were bent.

If your species of choice has a low number in the "Good Bends" column, that doesn't mean the wood is impossible to steam-bend. It only means that pine is a lot harder to bend than oak. With that in mind, perhaps you should consider either using a different wood species or switching to bent lamination to make the part.

Species	Unusable Bends	Marginal Bends	Good Bends
Alder	7	3	0
Ash	0	2	8
Domestic Cherry	3	4	3
Hard Maple	3	3	4
Hickory (Pecan)	0	2	8
"Honduran" Mahogany	8	2	0
Pine	9	1	0
Red Oak	0	1	9
Soft Maple	3	2	5
White Oak	0	2	8

UNUSABLE BENDS

These bends simply don't work well. Either there was excessive cracking on the convex face or severe crumpling on the concave face (most common failure); sometimes there might have been both. No amount of milling could save the part.

MARGINAL BENDS

These bends had some cracking on the convex face and a lot of crumpling on the concave face and as well. Heavy milling, removing lots of material, would remove most of the bending flaws.

GOOD BENDS

These bends had little or no cracking on the convex face and only light crumpling on the concave face. Light milling would remove any bending flaws.

Why do bends fail?

A. If the outside (convex) part of the bend stretches too much the piece will break. Tension pulls the fibers apart. This is the most common failure, regardless of species or steam-bending technique.

B. Next in frequency of bending failure comes species and stock selection. By now experimenting with bending samples of different woods has you (hope-fully) rethinking the notion of making that chair back out of Bubinga in favor of white oak.

C. Flaws in the piece being bent will almost invariably cause a failure. A small knot will harden a small area in the piece and render it almost impossible to bend.

D. Grain runout is where the grain of the wood does not run parallel with the board. It's easy to spot by sighting the reflection on the edge of the board. Runout will allow a tear to start right at the corner of the piece, forming a very nasty splinter. Surprisingly, the orientation of the grain in relation to the bend makes no difference.

E. Insufficient heat, box temperature that's too low, not enough time heating the piece, or perhaps taking too long to make the actual bend are sure ways to splinter your nice stock into firewood. If the heat doesn't get to the actual core of the piece or you're too slow, there's no way the bend will work.

Building a steambox

Once you've determined the size of the pieces you need to bend, build the steambox to fit that size. The smaller the box, the smaller the boiler you'll need to heat it. One way or another, you'll need to heat the box to at least 212°F and keep it there.

1. **Cut ¾-in. exterior plywood** or #2 pine into the top, bottom, sides, and ends of the box. If I have a longer box to make, over 8 ft., I normally use pine, as longer lengths are usually readily available. Use coated deck screws in assembly instead of steel nails or screws and drill holes in the sides for fitting in dowels.

2. **Fit in dowels** for shelving.

3. **Cut a piece of plywood** or pine for the door. Cut it to the right size to stay in place with a friction fit.

4. **Drill a hole** in the back of the box for the hose connection. Don't forget to drill vent holes to prevent build up of steam pressure. Make an extra vent hole to mount a meat thermometer.

Trial bending

1. **Cut a variety of pieces** all the same length and beginning at about ½ sq. in. up to ¾ sq. in. Try whatever species you have around the shop. Label them to keep track.

2. **Load your trial bending wood** pieces into the box, spreading them out so steam can circulate around them (in this photo, the lid has been removed for clarity, but normally the box will be complete by this step).

3. **Heat the box** and measure the temperature with a meat thermometer. The correct temperature should be 212°F. Write down the amount of time it takes for the box to reach full temperature. Heat the samples for about one hour per inch of thickness (30 minutes for a ½-in.-thick piece). This follows the general guideline to heat the pieces one hour per inch of thickness. Insulate the box with rigid foam if it is close to but not quite reaching 212°F.

4. **Take out a piece** (while wearing gloves) and shut the door. Try bending it slowly, then more quickly, around a gentle radius, then around a sharper one. Note how well or how poorly it bends. Repeat for all remaining trial pieces.

❯ warning

Live steam is not always visible and is very dangerous. Always wear gloves that will protect you at least to your forearm.

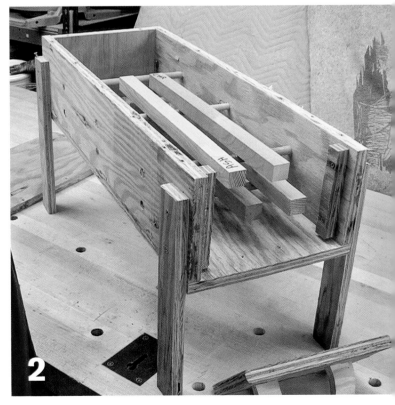

3

4

Bending solid wood around a form

In trial bending, you learned that most, if not all, of the wood you were bending freehand around the form broke. Without compression, the wood fibers were stretching to their breaking point. Bending solid wood around a form using a compression strap compacts the wood fibers, drastically improving your chances of bending without breakage.

1. **Quickly set the piece** between the form and the compression strap.

2. **Clamp the piece of wood** to the metal compression strap (see Chapter 7 for more on compression straps). The compression strap limits the amount the piece can stretch, making the piece compress instead.

3. **Start bending the piece** around the form. Make sure to over-bend the piece so that it has room to spring back into the shape you want.

4. **Use clamps and clamping blocks** to hold the piece to the form as the bend progresses.

5. **Clamp and leave the piece,** once it is successfully bent around the form, for 10 to 15 minutes.

6. **Undo the clamps and remove the piece** from the bending form. Place the piece on the cooling form and clamp it into place.

7. **Place the piece on the cooling form** on a rack and cover with a blanket over a heater. The wood must dry for 24 hours before it is set at its desired shape.

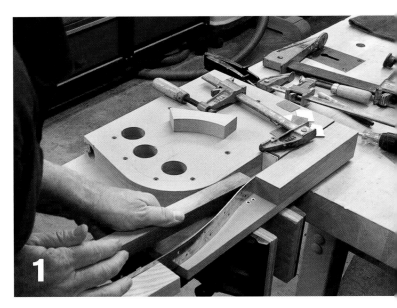

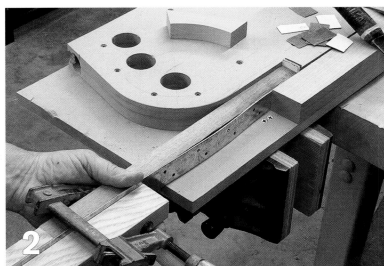

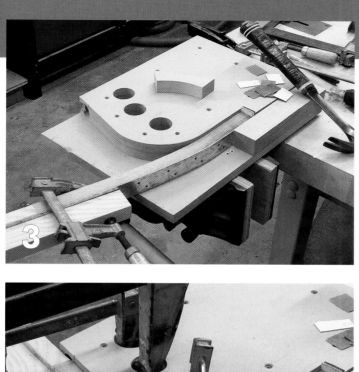

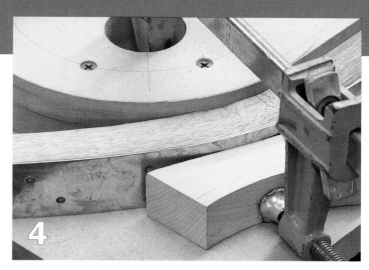

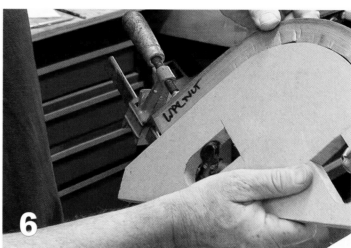

7

Steam-bending

I n the previous chapter, you discovered how difficult it was to bend even a ½-in. square of solid wood without breaking it. So how do you bend a thicker piece like a 1-in. square? This chapter is all about the key to steam-bending successfully: a compression strap. A compression strap can turn a seemingly impossible solid-wood bend into one that's possible.

If you let the piece bend naturally, it will stretch until it breaks. Using the compression strap, you're limiting the amount the wood can stretch. The piece has no choice but to compress. If you can compress the piece, you can bend it.

This shop-built compression strap is quick to make, costs nearly nothing, and really does the job.

How compression works

In order for something (not just wood) to bend, either the outside (convex) part of the piece has to get longer or the inside (concave) part has to get shorter—or some combination of the two.

Let's use the example of bending the back for a chair. We start with a board 1 in. square by 48 in. long, which will be bent around a 6-in. radius to form a 180-degree curve. After the bend, the inside of the curve can no longer be 48 in. It has to be more like 45 in. Where did 3 in. go? The wood actually got shorter. However, on the outside of the board, the wood had to stretch.

As you discovered when you were experimenting with your trial bends, breakage will occur on the outside of the bend. When you bend a chair back, you have to limit how much the outside of piece will stretch or it will break. So the key to bending successfully is to get the inside (concave) face of the bend to compress. Heating the wood softens it just enough to allow the wood to compress, which allows the board to bend. The key to bending without breakage is limiting the amount of stretch on the outside surface and controlling the compression on the inside surface.

What is a compression strap?

In its simplest form, a compression strap consists of a flexible metal strap with fixed blocks at each end. You apply the strap to the bending piece right out of the steambox. The strap stays in place on the convex face of the piece while you make the bend. As the piece bends, the end blocks apply pressure. The outside of the piece tries to stretch but is restricted by the strap. As the bend

Predicting compression

Some simple math illustrates the "stretch and shrink" necessary to bend. It's an easy way to figure the difference in length between the outer and inner surfaces of the bend. This difference has everything to do with predicting how difficult the bend will be.

For our example, we'll bend a 1-in.-square board 180 degrees around a 6-in. radius.

- For the inside circumference, multiply the inside diameter (12 in.) by Pi (3.1416), then divide by two to get half of the inside circumference, in this case, 18.85 in. (3.1416 × 12 in. ÷ 2 = 18.85 in.).

- Figure the outside circumference the same way. (3.1416 × 14 in. ÷ 2 = 21.99 in.). Half of the outside, or convex, circumference is 21.99 in.

- Now subtract the inside circumference from the outside (21.99 in. − 18.85 in. = 3.14 in.). There is a difference of 3.14 in. between the inside and outside of the chair back. If you don't believe it, take a look at the photos on the facing page. As much as I've thought about this over the years, I cannot avoid shaking my head that bends like this ever work—but they do.

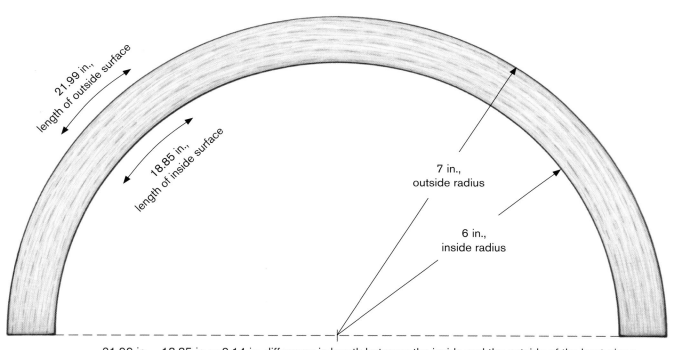

21.99 in., length of outside surface

18.85 in., length of inside surface

7 in., outside radius

6 in., inside radius

Not to scale

21.99 in. − 18.85 in. = 3.14-in. difference in length between the inside and the outside of the bent piece

Note the blue tape on the outside of the chair back and the green tape on the inside.

Now note the same pieces of tape removed from the bend and attached to the table, side by side. When measured, the green tape is approximately 3 in. shorter than the blue tape, thus showing the effects of compression when it comes to bending wood.

Compression
at work for bending

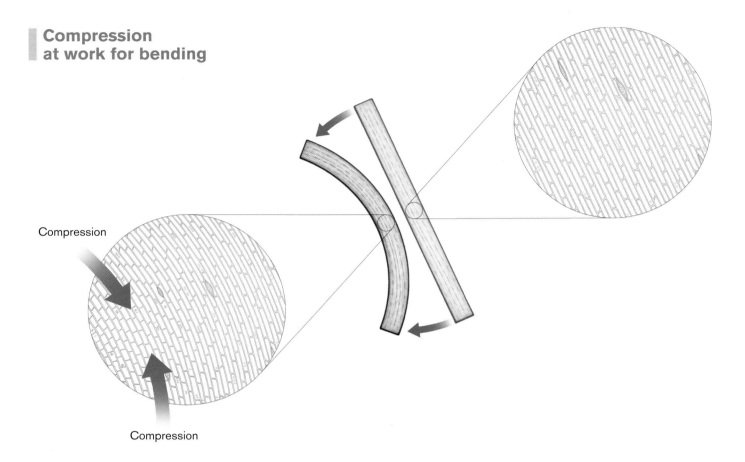

Compression

Compression

The essential compression strap

Without a compression strap, the outside of the bend usually stretches to the breaking point—and beyond. With a strap, the outside cannot stretch so the inside must compress.

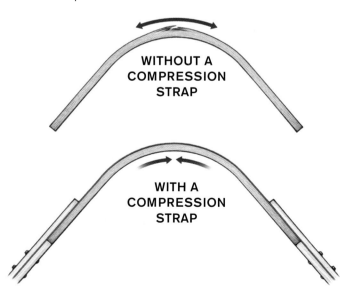

WITHOUT A COMPRESSION STRAP

WITH A COMPRESSION STRAP

progresses, there is more and more pressure on the end blocks. With the right setup and a little luck, it actually works.

The stronger the strap, the better

In Chapter 6, I suggested building a different steambox for every project. I offer the same advice for compression straps. Once you build the basic hardware, it isn't difficult to adapt it to different sizes. You can make it as simple or elaborate as you like as long as it's very strongly built. Your approach to this, like other setups, has a lot to do with volume. Are you making four chairs or four dozen? And because you can buy a good strap online, perhaps you'll decide not to make one at all.

If you build a compression strap, over-build it far beyond the strength factor you think you'll need. Even then, you'll probably break a few parts before you get it to really work. Don't underestimate the amount of end

It's nearly impossible to bend this ³⁄₄-in. piece around the form without a compression strap because the wood stretches and breaks.

This compression strap, created to fit shorter pieces of solid wood, is built into the form. The piece of wood shown here is thicker than the piece in the photos above and it bent around the form successfully because of the use of compression.

There is enough end pressure during the bending process to imprint the face of Thomas Jefferson in the end of the piece by placing a nickel between it and the end block.

pressure. It's easier to grasp the principal if you visualize the amount of pressure it would take to shorten (compress) our 1-in.-square board by several inches. Try placing a nickel between the end block and the end of the piece you're bending. Look closely and you'll be able to see Jefferson's profile imprinted on the end of the piece. It's difficult to imagine how much pressure these end blocks have to withstand until you try a bend or two.

Components of a compression strap

Start with the piece you're bending, then build the strap for the length, thickness, and width of the piece. As shown in Chapter 1, draw all of the parts full scale in three views (top, side, end) before you build. The end blocks go on one side of the strap, where the steamed piece will go. The handles with extended back blocks go on the other side of the strap, making a sandwich (see the drawing on the facing page).

Handles with backer blocks

The handles should extend beyond the length of the piece being bent to provide more leverage for the bend. The longer the handles, the less effort it will take to bend your piece. The length of the handles should be adjusted according to the severity of the bend. Start with 24-in. handles. If resistance is high, try longer handles up to a 36-in. maximum.

The handles should overlap the piece to be bent. They should extend over the bending strap itself to form backer blacks. On the inside of the handle, end blocks are bolted through the strap into the handle. These will limit the amount the piece can stretch. Without extended backer blocks, there is so much pressure on the end blocks that they'll rotate out of position and allow the piece to pop out of the strap (see "The importance of backer blocks" on the facing page).

Centering blocks

If the bend is a long one, say over about 3 ft. in length, the strap will also need centering blocks (see the photo on p. 86) to keep it centered on the piece being bent. A centering block every 18 in. will effectively hold the strap to the workpiece.

Metal strap

For the metal strap itself, start with the galvanized strapping pieces, called "coiled strap ties" used in house framing. Coiled strap looks a lot like plumber's tape but is much heavier gauge. The material is sold at home centers and lumberyards. It comes in various thicknesses, widths, and lengths. Being galvanized, it's better than plain steel because it doesn't stain heated wood, especially oak, as badly.

A very good alternative is a strip of stainless steel sheet metal. It's difficult to drill but works great because it won't stain the wood you're bending at all. A metalwork supply house will be able to cut it to just about any width and length.

The metal strap should be at least a foot longer than the bending piece. This allows plenty of length for bolting on the end blocks. Mark the holes in the strap. Then use a center punch on the hole locations to help guide the drill

Compression strap components

A compression strap is a pretty straightforward project. All you really need for the metalwork is a hacksaw and some twist drills. Everything you need is at your local home center.

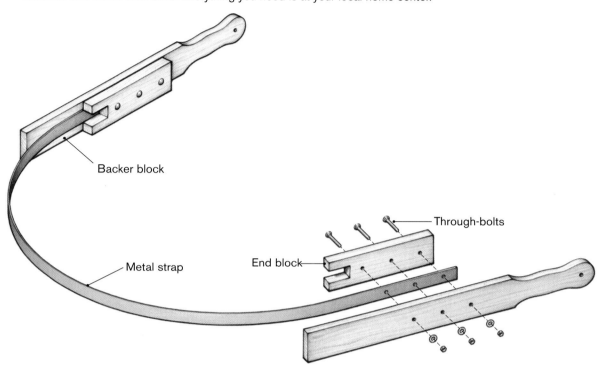

Backer block

Through-bolts

Metal strap

End block

The importance of backer blocks

These backer blocks need to extend along the bend. They keep the end blocks from rotating outward and allowing the piece to pop out of the strap.

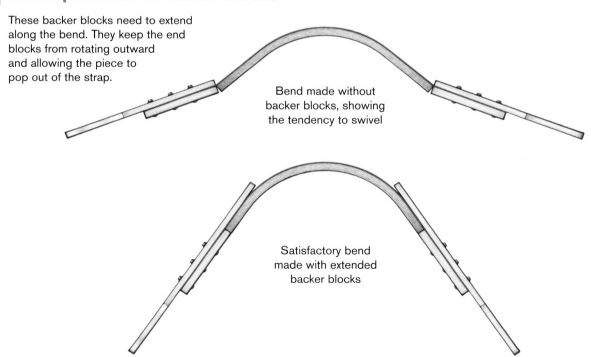

Bend made without backer blocks, showing the tendency to swivel

Satisfactory bend made with extended backer blocks

This centering block keeps the strap centered on the bending piece.

bit. Usually I drill the holes for the first bolt, then insert the bolt into the hole to tie the pieces together. Then using a twist drill bit intended for drilling metal I drill right through wood and steel strap for the rest of the bolts. Use a minimum of three ¼-in. bolts to hold the end blocks in place. Carriage bolts work well.

Commercially available compression straps

Lee Valley® makes the only commercially available small shop strap system I know of. Because the strapping material itself is spring steel it will last a lot longer than the shop-built strap described previously.

A major plus of this system is having an adjustable screw mechanism built into the end stop used to apply and/or release the end pressure. As you gain experience,

you'll know when you've applied enough end pressure and when you can afford to release some of it. For example, when you go beyond about 90 degrees of bend, the end pressure builds to the point where the inside of the piece tends to wrinkle excessively. Gradually unscrew the end block, thereby lessening the pressure as the bend progresses.

Bending a chair back

Before you bend the stock for your project, make a test run with red oak, which bends and compresses quite well. The idea is to at least test your setup. If you have trouble bending oak, you have some adjustments to make. If oak bends without trouble, then heat up the stock you've chosen for your project and give it a try.

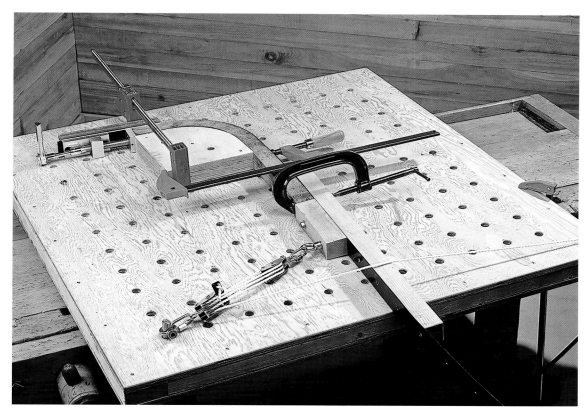

The Lee Valley bending strap is the perfect size and scale for bending small furniture parts. Featuring an adjustable end block and heavy spring-steel strapping material, this strap may be the only one you'll ever need.

Some wrinkling is unavoidable with a challenging bend, but the amount can be controlled by adjusting the end stop. But if you back off the end stop enough to avoid wrinkling the inside of the curve, you may allow the board to stretch too much and tear on the outside of the curve. Getting it just right is a balancing act.

When bending a batch of parts, place boards in the steam at 15-minute intervals. Numbers on the ends indicate the sequence.

Cutting the boards

If you cut lots of blanks it won't matter much if you mess up a couple of them. Cut them slightly oversize on the tablesaw then plane them down on all sides to final size. Last, cut them all to length.

Now's a good time to sight down the surface reflection to check for grain runout. Choose the blanks with the most runout to warm up. These will be more likely to fail. Save the nice straight grain pieces for actual workpieces for the project.

workSmart

Bending something like a chair back is not going to work perfectly the first time—or every time, for that matter. For example, if you plan to build four chairs, take the pressure off and mill extra blanks in case some of your bends fail. The worst that can happen is you'll have enough nicely bent blanks for an extra chair or two.

Heating the boards

When you place the boards in the steam, opening the box makes the temperature drop. Wait until the box is back to 212°F before noting the time. Then it's one hour in the steam per inch of board thickness.

Bending the boards

As the boards come out of the steam and are set into the compression strap, apply just a bit of end pressure. If you have the screw-type strap, get the end block just snug. If you're using shims, use all you can get in by hand. Then—as fast as you can—get the board onto the form.

The bending itself has to go quickly but carefully. Just as you experienced with the trial bends, the board needs some time to get used to the idea of bending. Fumbling with all the clamps will take some time, so go as quickly as you can without rushing the process.

You'll know right away if the piece cools too much to bend because you'll feel its resistance. Don't force it. If you do, chances are you'll break your compression strap. Take the setup off of the form, remove the piece from the compression strap, and throw it on the firewood pile. Then try again. Each time, the process will go just a little more smoothly.

It's a race to get to the beginning of the bend. The piece cools very quickly so there's no time to waste. But from this point on, it's go as fast as you can but as slowly as you need to get everything in alignment and adjusted. As you practice you'll discover the correct pacing for the process.

The setting form will hold the shape while the bend cools, dries, and sets.

Setting the bend

If you leave the board on the form for 5 or 10 minutes, that should be plenty of time for it to cool. Back off the tension screw or pull out the shims, pop the bend out, and immediately clamp the board to the setting form. The springback will be pronounced at this point, but the part easily should bend around the drying form.

Building a compression strap

1

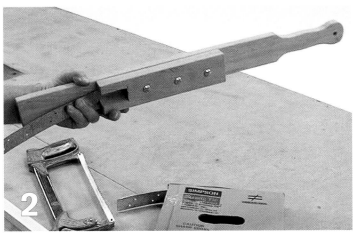

2

3

If you have just a few parts to bend, a shop-built compression strap is easy to make. You'll be working with metal, so you'll need a hacksaw and some twist-type, metalworking drill bits.

1. **Use metal strapping** material, called coiled strap ties. This material is available at lumberyards and home centers. The strap will only last for a few bends, but it's inexpensive, easy to work with, and readily available.

2. **Make the end block** to fit the part you're bending. It's pretty easy to cut it out on a bandsaw. If it's the same thickness as the part you're bending, it works well when you clamp the bend to the form.

3. **Extend the handle** along the strap to create a backer block, which keeps the end block in line.

4. **Drill through the handle,** the end block, and the strap. Use carriage bolts to hold it all together. Bolt one end and cut the strap to the length of the piece. Place the piece you're going to bend in place, position the parts for the other end, and then drill the bolt holes for the other end of the strap.

5. **Bolt the other end in place** and the strap is complete. With each use the metal will stretch. When it gets too long or distorted, simply cut a new piece of steel strap to length and replace.

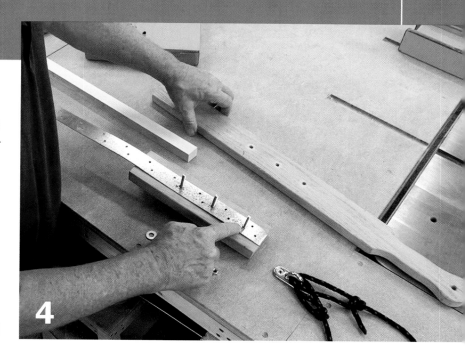

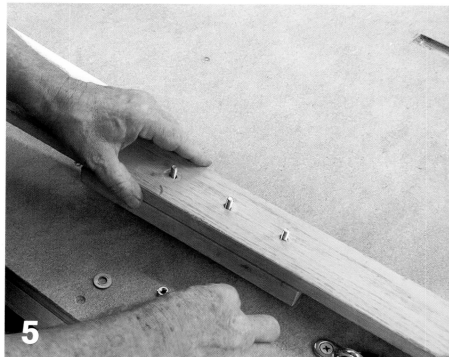

Bending a large part with a compression strap

It's possible to bend large parts single-handedly. It's admittedly easier with a helper or two, but worth noting that with the right setup, you can bend them on your own. The challenge is to bend the part before it cools.

1. **Secure the form to the workbench.** Compression creates a lot of pressure, so be sure your form is attached well enough to stay put during the bend.

2. **Locate the center mark** on the piece so you can quickly get your setup aligned and clamped to the form. A bar clamp is good to use because it's very fast. Quickly hook up the pulleys and take up the slack to get the bend started.

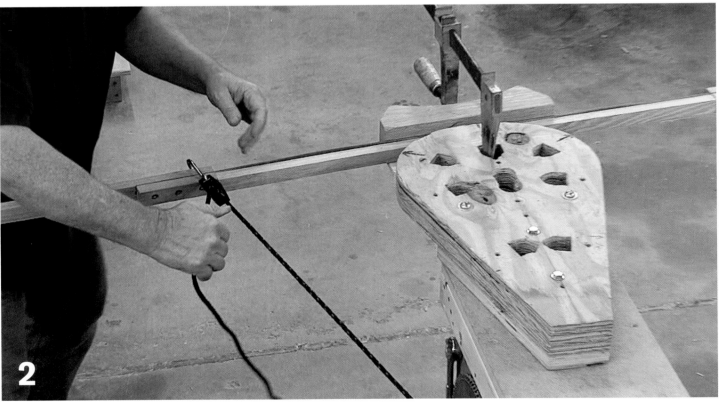

3. Using pulleys, go back and forth from one side to the other, pulling the piece around the form like a big crossbow. The pulleys make it easy to do the actual bending. You'll reach a point where the direction of the pull sort of bottoms out.

4. Use bar clamps to pull the bend in the rest of the way. Note how much over-bend is used for this chair back. The idea is to allow the piece to relax to its final shape.

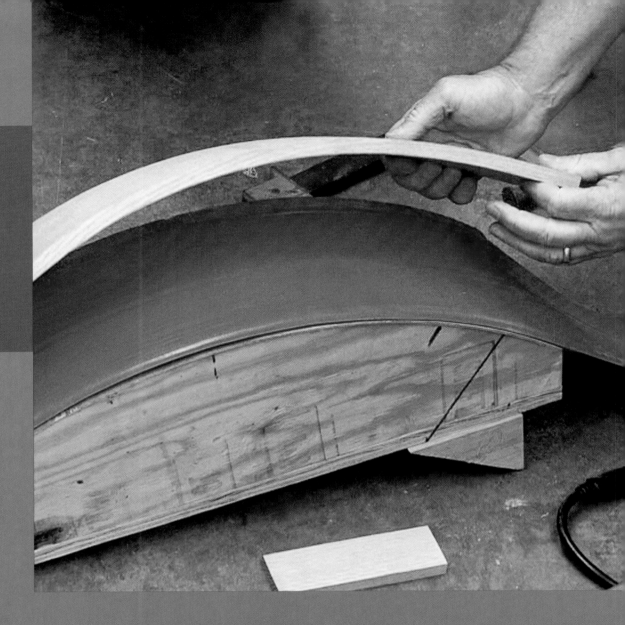

Bending with a heating blanket

The technique of bending wood with heat alone hasn't changed much for centuries, going back to when luthiers would use a hot stovepipe to bend the sides of stringed instruments. These days, there are more user-friendly sources of heat, including electric heating pads, heat guns, and hot pipe setups. But the technique remains the same: heat the wood, bend slowly, and repeat as necessary.

In this chapter, we will focus on using a heating blanket to bend wood with dry heat. The hot-blanket method will work for pieces up to about ½ in.

How dry-heat bending works

Think of wood, when it's still part of a living tree, as cellulose tubes held together with natural glue called lignin. These tubes are filled with water or sap when the tree is still growing. As soon as the tree is cut, this liquid begins to evaporate. What's left are the cellulose tubes (wood fibers) held together by the lignin.

Think of lignin as something like hot-melt glue. Like hot-melt glue, it is not particularly water soluble, but when heated, it softens just a little. It hardens again as it cools. When this softening takes place the wood is said to be plasticized; in other words, the wood is made slightly more flexible when heated.

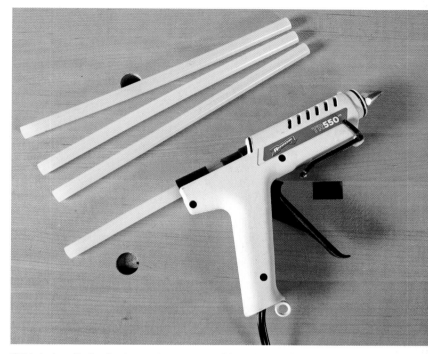

If this hot-melt glue is at room temperature, it's solid and relatively hard, like the lignin in wood.

When heated, the hot-melt glue becomes liquid, like lignin.

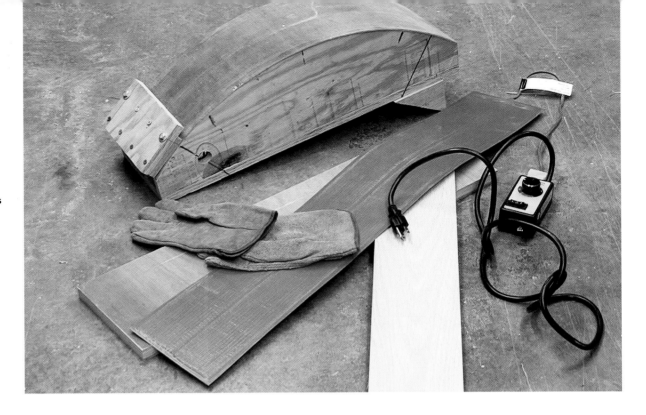

This amazing invention was developed to keep spacecraft components warm. Little did the developers know that wood benders would be using these heating blankets to make guitars and chair parts.

workSmart

I've found that heating the wood for a while flat—that is to say, before trying to bend it—lets the wood soften a bit. Before you start, wet the wood on the side facing the heat to help keep it from scorching. Some instrument makers layer pieces of saturated cloth or paper on the surface of the wood to help keep it moist during the bend. This will generate a small amount of steam, which will help heat the wood and also help prevent scorching.

Remember it's the compression

It's compression that will allow a piece to bend without breaking. If the piece of wood stretches too far, it will break. You'll want to apply heat on the inner (concave) side of the bend. This allows the wood to deform (compress on one side) and bend successfully.

Bending with a silicone heating blanket

The new, high-tech silicone heating blankets are right in there with Velcro®; a relatively new innovation developed to keep spacecraft machinery warm. This isn't the heating pad you use when your back is sore. These heating blankets are more like a piece of rubber that gets hot—really hot—well above the temperature that will scorch or even light wood on fire.

The blanket itself bends right along with the piece. This means it supplies heat during the actual bending process. Unlike steam-bending, the piece never has a chance to cool down and stiffen.

Bending with the blanket is tricky because it gets too hot to touch so it's difficult to handle. I use gloves, but even then it's very hot. After the piece is heated as much as it's going to be while laying flat, place it on the form and bend very slowly until it's the right shape.

Bending solid wood with a heating blanket

This heating blanket is flexible and very strong, with a working temperature as high as 400°F to 500° F. A temperature control is vital. Without one, the blanket will get way too hot and might even cause the wood to catch fire, so set the control to halfway between medium and high. This setup works fine with pieces up to about ½ in. thick and bent to a very slight radius. This example is red oak, ½ in. thick by 6 in. wide, bent to a shape suitable for a chair crest rail.

1. **Preheat the piece** by laying it on the heating blanket. Make a kind of sandwich, with the bending piece on top, the blanket in the middle, and a flat board below.

2. **Turn on the heating blanket.** Let the wood heat for about half an hour before trying to bend it. As with any other heat-bending method, it takes a while for the lignin to get warm enough to soften sufficiently.

3. **Place the heating blanket,** then the piece you're bending, on top of the form. Always wear gloves when handling these extremely hot items.

(continued on p. 98)

warning

Wearing gloves is a must when handling the heating blanket. Sometimes the blanket gets so hot that even the gloves don't provide enough protection. Use your best judgment when handling hot materials.

4. **Allow the piece to bend** gradually on its own. Once it does, gently apply pressure in increments around the curve. When the lignin is warm enough to give, the piece will bend quite easily. Take it slow to give the piece time to readjust to its new shape.

5. **Once the piece has bent** to nearly the finish shape, remove it and quickly turn it end for end and then reapply bending pressure. This will even out the shape of the piece. Apply clamps to the middle and end of the piece. Once the piece is bent, turn off the heating blanket, let cool overnight, take off the clamps, and remove the bent piece from the form.

metric conversion chart

Inches	Centimeters	Millimeters	Inches	Centimeters	Millimeters
1/8	0.3	3	13	33.0	330
1/4	0.6	6	14	35.6	356
3/8	1.0	10	15	38.1	381
1/2	1.3	13	16	40.6	406
5/8	1.6	16	17	43.2	432
3/4	1.9	19	18	45.7	457
7/8	2.2	22	19	48.3	483
1	2.5	25	20	50.8	508
1 1/4	3.2	32	21	53.3	533
1 1/2	3.8	38	22	55.9	559
1 3/4	4.4	44	23	58.4	584
2	5.1	51	24	61.0	610
2 1/2	6.4	64	25	63.5	635
3	7.6	76	26	66.0	660
3 1/2	8.9	89	27	68.6	686
4	10.2	102	28	71.1	711
4 1/2	11.4	114	29	73.7	737
5	12.7	127	30	76.2	762
6	15.2	152	31	78.7	787
7	17.8	178	32	81.3	813
8	20.3	203	33	83.8	838
9	22.9	229	34	86.4	864
10	25.4	254	35	88.9	889
11	27.9	279	36	91.4	914
12 1/2	30.5	305			

resources

Clamp sources

- **Jorgensen® Clamps**
  www.adjustableclamp.com

Glue sources

- **DAP Plastic Resin Glue**
  www.dap.com

- **Unibond 800 glue**
  www.vacupress.com

- **West System® Epoxy**
  www.westsystem.com

- **System Three® Epoxy**
  www.systemthree.com

Lubricants

- **Top Coat Lubricants**
  www.rockler.com

Veneer sources

- **Certainly Wood®**
  www.certainlywood.com

- **Constantines**
  www.constantines.com

Bending strap hardware

- **Lee Valley**
  www.leevalley.com

Web site for bending information

- **Lon Schleining's Web site**
  www.woodbender.com

Bending blankets

- **Blues Creek Luthiers Suppliers**
  www.bluescreekguitars.com

Magazines

- ***Wooden Boat* magazine**
  www.woodenboat.com

- ***Fine Woodworking* magazine**
  www.finewoodworking.com

Books

- ***The Complete Manual of Wood Bending* by Lon Schleining**
  (Linden Publishing, 2002)

- ***Tage Frid Teaches Woodworking* by Tage Frid**
  (Taunton Press, 1981)

- ***Wood Bending Handbook* by** Stevens and Turner (Fox Chapel Publishing, 2007)

- ***Wood Bender's Handbook* by** Zachary Taylor (Sterling Publishing, 2008)

- ***Make a Chair from a Tree* by** John Alexander (Astragal Press, 1994)

- ***Understanding Wood* by** Bruce Hoadley (Taunton Press, 2000)

glossary

Backer blocks

These blocks keep the compression strap end blocks from rotating out of position when bending.

Bar clamp

Also known as an F-clamp or speed clamp. There are two kinds, with and without clutches.

Batten

A thin strip of solid wood tacked to a wooden blank for pattern-making. More generally, it's a piece of wood used to draw curves.

Bending form

The bending form is the structure around which the bend is made. It often has clamp pockets and guide blocks to facilitate the bend.

Bent lamination

The process of bending thin layers of wood over a form. Gluing them together keeps their shape once the clamps are released.

Boiler

A vessel in which water is heated to produce the steam that heats the steambox.

BTU

British Thermal Unit: A unit of energy; specifically the energy it takes to heat one pound of water one degree Fahrenheit, or roughly the heat output of a single match.

Bundle

A glued-up stack of laminates ready for the bending form.

Centering blocks

"U" shaped blocks that keep the compression strap centered on the wood being bent during the actual bend.

Circumference

The length around the perimeter of a circle.

Clamp blocks

Segments of the female bending form cut into separate blocks. The blocks distribute the clamp pressure.

Cleat

A piece fastened to or across something to give it strength or hold it in position. For example, a cleat is bolted to the bending form so it can be clamped in the vise.

Compression

A force that decreases the volume of an object by squeezing and shortening it. Compression is applied to a bend with the compression strap to keep the wood from stretching and breaking.

Compression strap

A strap that applies compression by using end blocks to limit the amount of lengthening possible when a piece of solid wood bends.

Cooling form

A form used to hold the piece in exact shape while it cools and dries.

Concave

An inward (female) curve.

Convex

An outward (male) curve.

Curing time

The amount of time it takes for glue to reach its full strength.

Dowels

Solid, cylindrical, wooden rods used in a steambox as shelf support for wooden pieces being heated.

Drop cloth

A piece of cloth laid over the floor during the glue-up and clamping stages of bent lamination.

Edge-clamping

Vertically clamping the sides of a bent lamination to ensure the laminates are properly aligned on the form and with one another.

End grain

The open grain showing at the end of a board.

Flitch

A stack of veneers matched like they were before they were sawn from a solid block.

Glue creep or cold creep

A charactistic of flexible glues, like yellow woodworker's glue, allowing pieces under tension to slide, or creep past one another over time. In bent lamination, if flexible glue is used (not recommended), glue creep occurs as the bend tries to straighten out.

Grain runout

Where grain of the wood does not run parallel with the board. Grain runout can be seen in the reflection on the surface of a board.

Guide blocks

Blocks that keep the bent lamination aligned on the form. They also act as shelves to hold the clamp blocks steady while clamping up the bundle of laminates.

Hook scraper

The recommended tool for removing dried glue from wooden surfaces. It looks like a regular paint scraper sharpened with a file.

Kerf

The width of a sawcut or the amount of wood removed with the saw blade.

Laminates

Thin strips of wood that make a bent lamination when glued together and clamped to a form.

Lignin

A substance that fuses with cellulose within the cell walls of woody plants to create the rigidity necessary to hold them erect. Think of it as glue that softens just a bit when it's heated.

Live steam

This form of steam is invisible. White vapor is condensed live steam. The invisible form is the most dangerous because of its high temperature.

Male bending form

A convex form around which to bend wood.

MDF

Medium-density fiberboard

Milling

Squaring and straightening the bent lamination.

Moisture content (MC)

The percentage of moisture in a piece of wood by weight.

Outfeed table

A device to support a board while it's being sawn with a tablesaw.

Part

An element that will ultimately be a part of a finished furniture piece such as a chair back.

Portable surface planer

The portable or benchtop version of a machine that planes boards to an even thickness and flattens their surface.

Radius

A line segment from the center of a circle to its perimeter.

Silicone heating blanket

A space-age product developed to keep fuel and other liquids warm in space. Electricity heats the blanket and the blanket applies the heat needed to bend wood.

Springback

A condition that occurs when a bent part is released from the bending form and partially straightens out due the wood's elasticity.

Steam-bending

The process of using steam heat to soften the lignin in the wood and make it malleable enough to bend.

Steambox

The box in which wood is heated for steam-bending.

Stranded packing tape

This very strong tape is characterized by its lengthwise fibers that give it very high tensile strength.

Thickness jig

A jig for setting the width of laminates so that they can be cut on waste side of the blade of a tablesaw. This allows the larger part of the board to be between the blade and the rip fence.

Vent holes

Holes made deliberately in the steam box to allow steam to circulate and provide access for a thermometer.

index